Farming System

Principles and Practices

Farming System

Principles and Practices

Authors :

C. JAYANTHI

P. DEVASENAPATHY

C. VENNILA

SATISH SERIAL PUBLISHING HOUSE

403, Express Tower, Commercial Complex, Azadpur, Delhi-110033 (India)
Phone : 011-27672469, 27672852, Fax : 91-11-27672046
E-mail :info@satishserial.com, hkjain1975@yahoo.com
Website : www.satishserial.com

Published by :

SATISH SERIAL PUBLISHING HOUSE

403, Express Tower, Commercial Complex, Azadpur, Delhi-110033 (INDIA)
Phone : 011-27672852 Fax : 91-11-27672046
E-mail : info@satishserial.com, hkjain1975@yahoo.com

Reprint, 2019

ISBN 81-89304-49-6

Composed, Designed & Printed in India

Tamil Nadu Agricultural University

Coimbatore-641 003
Tamil Nadu, India

Prof. C.Ramasamy
Vice-Chancellor

FOREWORD

Population in most tropical countries is increasing by leaps and bounds, necessitating these countries to produce more food from the limited cultivable lands available. The Agricultural systems have encountered various kinds of ecological and environmental problems due to the growing pressure of human population on the carrying capacity of land and the productivity. While the ultimate goal of Sustainable Agriculture is to conserve the natural resource base, protect the environment, enhance the health and safety of human population over a longer period, this can be achieved by seeking optimal use of internal and external input mix in a way that provides acceptable levels of sustainable crop and livestock productivity resulting in economically profitable returns to the farmers and sufficient food commodities to meet the requirements of the Nation.

There occurs a need to increase the food, feed and fodder requirements to meet the demand of increasing population with the available limited resources. To have profitable returns from the farm enterprise, it is necessary to link various enterprises such as crop, livestock, forage *etc*. So, understanding the scientific basis of cropping system, forage and its importance, livestock maintenance and linkage and farming system approach is essential.

Farming system therefore, assumes greater importance for sound management of farm resources to enhance the farm productivity, reduce the environmental degradation, and improves the quality of life of resource poor fanners and to maintain sustainability. Farming System not only supplements the income of farmers, but also provides an opportunity to solve the malnutrition problem that exists in the diet of the Indian farmers.

Many researchers have developed science based sustainable integration of agricultural systems for adoption by the farmers in Tamil Nadu. Comprehensive information about the farming system, its principles and practices is seldom available. In this context, the effort made by the authors to prepare the manuscript is commendable. This will definitely serve as a source book for the Post Graduate Scholars and Research workers. I congratulate the authors, for the efforts taken in bringing out the book and wish them success in all their endeavors.

Coimbatore
August 14, 2007

C. RAMASAMY

PREFACE

The population in most tropical countries is increasing leaps and bounds, warranting more production of food from limited cultivated lands. Many challenges are faced in our quest to achieve sustainable food security with shrinking land resources for agriculture to produce additional food grains to meet the requirement of the prognosticated population in our country. Furthermore, as agriculture becomes more intensive, soil and waste degradation threaten the sustainable agricultural production. India would need to produce additional food grains of 100 and 160 million tonnes by 2030 and 2050 AD, respectively, to feed its projected population.

Compounding the problems arising from the growing pressure of human population on the carrying capacity of the land and productivity, the agricultural systems have met with various kinds of ecological and environmental problems. The ultimate goal of sustainable agriculture is to conserve the natural resource base, protect the environment, and enhance the health and safety of human population over a longer period. This can be achieved by seeking the optimal use of internal production inputs in a way that provide acceptable levels of sustainable crop productivity and livestock production resulting in economically profitable return.

Low external input usage in farming would be achieved through multiple cropping with diversified activities like livestock keeping, mushroom cultivation, bee keeping, agroforestry, fishery, sericulture, forage crop production *etc.* (Farming systems). The demand must be met from limited resources such as land, water, labour and energy. Farming systems have always been to promote agricultural change. It views the whole farm as a system with the integration of crops, animals, soil, other inputs and environmental influence wherein the farm family attempts to produce outputs within the limitations of its capability and resources and the socio-cultural setting. The ancient combination of livestock and crop activities had helped farmers in the past, almost all over the world, to use the manure as fertilizer for crops, and the crop residues as feed for livestock. However, there had been losses in the use of farm adhoc practices.

Adequate supply of nutritious feed and forage is a must to meet the food requirements of animals. So, there is a need to increase the forage production. A comprehensive idea on forage production and preservation will help in increasing the productivity of forage to meet the nutritional requirement of animals.

The Farming System has revolutionized conventional farming in some countries, especially in tropical and sub tropical regions. Science based sustainable integration of agricultural systems that would address the aspirations of the farmers and the concerns of the society were developed by researchers.

In this context, understanding the cropping system and management, forage crops and its importance, allied enterprises and its production and farming system approaches is necessary, which will help in increasing the productivity, maintaining soil health and to protect the environment. As a whole, the book will provide comprehensive information on the above subjects and definitely serve as a reference book for the agricultural students, extension personnel, scientists and policy makers.

The authors profoundly thank Dr. C. Ramasamy, Vice-Chancellor, Tamil Nadu Agricultural University, Coimbatore for his encouragement and suggestions for writing the book, and providing foreword.

The authors acknowledge Dr. K. Vanangamudi, Dean (Agriculture), Tamil Nadu Agricultural University, Coimbatore for his constant encouragement and critical review in preparing this book.

The authors express deep indebtness to Deputy Director General (Education), ICAR, New Delhi for providing financial assistance through ICAR Development Grant for preparing this book.

Due thanks are extended to all the scientists from whose work we have drawn the results for this text books.

Authors

CONTENTS

CHAPTER - II
FEED AND FORAGE CROPS

CHAPTER - III
FARMING SYSTEM

Introduction

Agriculture system today may be described as goal oriented manipulations of ecosystems for human gains. Yield and profit maximization approach has led to serious environmental, ecological, economic and social problems. Farm and rural population has steadily declined leading to an overall economic and social demise of many rural communities. Human society today is dominated by rapid technological and political innovations, summed by terms like globalization or information technology. The disseminated information is making more conscious on the quality of food and environmental consequences.

The population in most tropical countries is increasing leaps and bounds, warranting more production of food from limited cultivated lands. Many challenges are faced in our quest to achieve sustainable food security with shrinking land resources for agriculture to produce additional food grains to meet the requirement of the prognosticated population in our country. Furthermore, as agriculture becomes more intensive, soil and waste degradation threaten the sustainable agricultural production. India would need to produce additional food grains of 100 and 160 million tonnes by 2030 and 2050 AD, respectively, to feed its projected population.

Our production of food grains per unit area almost has to be doubled from what we are obtaining today. This could be made possible by putting the land, both irrigated and rainfed under intensive cultivation. Fortunately, most of our states lie in tropics and blessed with abundant solar energy thus making cropping possible round the year. The only way to increase agricultural production is to increase the productivity per unit area per unit time. In the scientific era of agriculture, cropping systems, genotypes, geometry of planting and management practices are designed to increase the productivity per unit area per unit time, simultaneously making efficient use of available resources and stabilizing yields.

The ultimate goal of sustainable agriculture is to conserve the natural resource base, protect the environment, and enhance the health and safety of human population over a longer period. This can be achieved by seeking the optimal use of internal production inputs in a way that provide acceptable levels of sustainable crop productivity and livestock production resulting in economically profitable return. The choice of components in the farming system and their management must ensure that the interaction between components should be the most complimentary with least competitiveness. The success of integrated agricultural allied enterprises with cropping for different farming situations has been studied by many researchers under different agro-eco systems.

Forage and feeds are major inputs in animal production which constitutes almost 70 per cent of total cost of production. Livestock contributes about 25 per cent of total income obtained through agriculture. Adequate supply of nutritious forage and feeds to the animals is a must to meet their nutritional requirement for maintenance and increased productivity. So it is necessary to increase the productivity of forage crops by planting suitable grasses, cereals, legumes and top feed tree species. In the last three decades, scientists and students at Tamil Nadu Agricultural University have conducted several research programmes addressing various issues and fruitful results were drawn on cropping system, forage production and integrated farming system. In this book, concepts, principles and agronomic management of cropping system, major forage crops, cultivation practices of forage crops and the farming system principle, methodology and research findings have been compiled for the use of undergraduate and post graduate students.

•••

Chapter 1

Cropping System

Agriculture system today may be described as goal oriented manipulations of ecosystems for human gains. Yield and profit maximization approach has led to serious environmental, ecological, economic and social problems. Farm and rural population has steadily declined leading to an overall economic and social demise of many rural communities. Human society today is dominated by rapid technological and political innovations, summed by terms like globalization or information technology. The disseminated information is making more conscious on the quality of food and environmental consequences. A cropping system adopted on a farm has more obvious and detectable social, ecological, economic and environmental implication. So, it is necessary to understand the principles and concepts underlying the entire ecosystem to have profitable, resource efficient and environmentally sound ecosystem.

1. PRINCIPLES AND CONCEPTS

For studying a system, as an entity made up of all its components and their interrelationships together with relationships between system and its environment

System approach hitherto extensively employed in business and industry is being introduced in crop production also with a view to use the available resources effectively and to increase the returns to the farmer. Accordingly crop production research has been reoriented as cropping system research in the country.

Agronomic research aims at optimizing input levels of such variable crop production factors as fertilizers and pesticides. Thus agronomic research increases the resource use efficiency for a given crop, cropping systems and cropping pattern (spatial or temporal combinations of crops in any one plot; generally time is limited to one year and a plot is defined as a contiguous area of crop planted in a homogenous manner during the defined period) as a variable with a view to use the available resources more efficiently. Hence, the objective of any cropping system is to increase the efficiency or utilization of resources - land, water and solar radiation. The efficiency is measured by the quantity of produce obtained per unit resource in a unit time. Productivity is measured in produce weight, calories or monetary units.

The productive base of a cropping system is the crop growth and yield. Crop yield ($\overline{Y}$) can then be considered a function ($\overline{f}$) of management factors ($\overline{M}$) and environment ($\overline{E}$)

$$\overline{Y} = f\,(\overline{M}, \overline{E}).$$

For the cropping systems researcher, management includes the type and arrangement of crops in time and space (cropping pattern), choice of variety, methods of stand establishment, pest management and harvest. Environment is composed of such land and climate related variables as rainfall, irrigation, soil, solar radiation and temperature and of availability of such resources as power, labour and cash (Harwood, 1975). Economic factors such as cost of inputs, price of produce, interest rates, etc., should be included in the environment components, if they cannot be altered for that time or should be included in the management term if the researcher is suggesting a change in the policy. Hence, it is evident that the management term is treated as a variable and the environment term as invariant. A cropping system researcher studies the interaction between *M* and *E* and seeks to determine how he should vary his cropping pattern, *M*, to optimize the returns for different production environments. In this concept, *E*, becomes a fixed constraint and the interaction between *E* and *M* gets merged with *M*. On this basis, cropping system can be defined as cropping pattern and its management to derive benefits from a given resource base under specific environmental condition. The term cropping system can be applied to a farm or a region (Sandstra, 1977).

The implication is that cropping system is location specific and to develop an alternate cropping system for a location, the prevailing environment of that location should be clearly understood. In suggesting an alternative cropping system for a location, it is generally assumed that the available physical resources are not fully exploited and hence by intensification of cropping, this lacuna can be removed. The crop intensification techniques include intercropping, relay cropping, sequential cropping and ratoon cropping.

The use of cropping system as a strategy for increasing productivity and the income of small farmers to

- Promotion of high yielding varieties of basic food crops
- Filling the protein gap
- War on waste
- Mobilization of human resources for rural development, and earning and conserving foreign exchange

Need

The cropping system research to date has adequately demonstrated the following potentials for adopting cropping system as a development strategy in contrast to monoculture

- Improved stability of food supply throughout the year
- Increased total food production per land unit per year, generally accompanied by an increase in total income for the farmer
- Improved distribution of income throughout the year
- Increased total employment of labour throughout the year, and
- Improved nutrition for the farm family from crop diversification

2. CROPPING SYSTEM - DEFINITION AND CLASSIFICATION

System

System means an arrangement of components which process input into output. Each system consists of boundaries, components, interaction between components, inputs and outputs.

Crop system

Crop system is an arrangement of crop population that transform solar energy, nutrients, water and other inputs into useful biomass. The crop can be of different species and variety, but they only constitute one crop system if they are managed as a single unit. The crop system is a sub-system of a cropping system. For example, in the maize crop system, maize is the dominant crop which is grown in association with other crops.

Cropping system

Cropping system is a land-use unit comprising soils, crop, weed, pathogen and subsystems that transform solar energy, water, nutrients, labour and other inputs into food, feed, fuel and fibre. The cropping system is a subsystem of the farming system.

The cropping patterns used on a farm and their interaction with farm resources, other farm enterprises, available technology and environment (physical, biological and socio-economic) which determine their make up, constitute the cropping system.

The objective of any cropping system is efficient utilization of all resources *viz.* land, water and solar radiation, maintaining stability in production and obtaining higher net returns. The efficiency is measured by the quantity of produce obtained per unit resource in a unit time.

The objective of traditional agriculture was to increase the production by two means

a. by increasing the productivity per unit area of the crop

b. by increasing the area of cultivation

But two more dimensions are added to modern agriculture

c. to increase the production per unit time

d. to increase the production per unit space

In the traditional cropping system, mixtures and rotation were developed by the farmers over years of experience by trial and error to suit specific ecological and socioeconomic conditions to attain yield stability, whereas modern scientific cropping system has three pillars.

A. *Genotype:* Genotype means genetic make up of seed

B. Geometry of planting

a. Shape of planting pattern on the land surface

b. Space of the area available for the individual plant.

Geometry of planting may be circular, rectangular, square type or cubical. It is indirectly related to plant population. Cubical pattern of planting has maximum plant population.

Plant population may be defined as

a. Size of area available to the individual plant

b. Number of plants per unit area

C. Management practices

Management practices include all the practices of crop production. For the cropping systems, management includes,

i. Type and arrangement of crops in time and space *i.e.* cropping pattern

ii. Choice of variety

iii. Method of stand establishment

iv. Pest management and harvest

Cropping pattern

The yearly sequence and spatial arrangement of crop or crops and fallow on a given area.

Farm household system

A group of usually related people who, individually or jointly, provide management, labour, capital, land and other inputs for the production of crops and livestock, and who consume at least part of the farm produce.

Difference between cropping pattern and cropping system

Cropping Pattern	Cropping System
Crop rotation practiced by a majority of farmers in a given area or locality.	Cropping pattern and its management to derive benefits from a given resource base under specific environmental condition.
Type and arrangement of crops in time and space.	The cropping pattern used on a farm and their interaction with farm resources, other farm enterprises and available technology which determine their make up.
Yearly sequence and spatial arrangement of crop or crops and fallow on a given area. The proportion of area under various crops at a point of time in a unit area	Pattern of crops taken up for a given piece of land, or order in which the crops are cultivated on a piece of land over a fixed period, associated with soil management practices such as tillage, manuring and irrigation

Sole cropping/Solid planting

It is defined as the cultivation of one crop variety alone in pure stand at normal density in a certain time and place.

Monoculture

Monoculture is the repetitive growing of the same crop on the same land. Monoculture is also called single cropping or mono-cropping. It is a system of growing the same crop on the same land year after year.

Multiple cropping

The intensification of cropping in time and space dimensions. Growing two or more crops on the same field in a year.

Shifting cultivation/Land rotation/Jhuming

Forest land is cleared and cultivated. Due to cultivation of the same crop on the same cleared forest land year after year; soil productivity is lost and the crop is shifted to other slashed and burnt land. Here same crop is grown year after year. In this case land is rotated but crop is fixed. Therefore, it may also be called land rotation. Shifting of land hence called shifting cultivation. Also called *Jhum* cultivation. It causes soil erosion. Mostly, practiced in North-Eastern states of India, Chhotanagpur Plateau of Jharkhand, M.P. and in hilly areas.

Crop rotation

The repetitive cultivation of an ordered succession of crops (or crops and fallow) on the same land. One cycle may take one or more years to complete.

Ley farming

Ahlgren (1956) defined it as a system of farming, in which grasses and legumes are included in proper rotation for hay, silage and pasture to care for maximum livestock needs and to improve and conserve the soil fertility.

Thus, ley farming is a dynamic and integral system of agriculture which encamps, soil, plants and livestock together. It denotes a system of farming whereby emphasis is placed on the value of grass-legume mixtures to get fodder for livestock and build up soil fertility for obtaining higher yields of subsequent arable crops.

Sustainable agriculture/Eco-farming

Sustainable agriculture is a concept to produce sufficient food to meet the needs of the present generation without eroding the ecological assets and productivity of life supporting systems of future generations.

Natural farming is an excellent illustration of sustainable agriculture. It is also known as ecological farming/ eco-farming or organic farming or permaculture. It is called eco farming because ecological balance is given importance since organic matter is the main source for nutrient management. It is a system of cultivation with use of manures, crop rotation and minimal tillage. Sustainable agriculture also involves agro-forestry, multi-level cultivation and integrated animal husbandry. The term sustainability denotes the characteristic of a process that can be maintained indefinitely and sustainable use of the ecosystem refers to making use of the system without impairing its capacity for renewal or regeneration.

Types of Agriculture	Objective	Quantity of Input	Quantity of Output
Subsistence Agriculture *i.e.* low level equilibrium	To sustain life and family need	Low	Low
Commercial Farming *i.e.* high level equilibrium	To obtain high Income	High	High
Sustainable Agriculture *i.e.* Natural equilibrium or ecological equilibrium	Ecological balance	Low	High

3. INTENSIVE CROPPING

Cropping system based on climate, soil and water availability have to be evolved for realizing the potential production levels through efficient use of available resources. The cropping system should provide enough food for the family, fodder to the cattle and generate sufficient cash income for domestic and cultivation expenses. These objectives could be achieved by adopting intensive cropping. Methods of intensive cropping include multiple cropping and intercropping. Intensive cropping may pose some practical difficulties such as shorter turn-around time lapse for land preparation before the succeeding crop and labour shortage at peak periods of agricultural activities. These practical handicaps can easily be overcome by making modifications in the cropping techniques. Alteration of crop geometry may help to accommodate intercrops without loosing the base crop population.

3.1 Classification of Cropping Systems

a) *Mono Cropping :* Cropping intensity is always 100%

b) *Multiple cropping :* Growing of two or more crops on same field in a year. It is the intensification of cropping in time and space dimensions *i.e.* more number of crops within a year and more number of crops on the same piece of land at any given period. The principal limiting factors in adoption of multiple cropping are low intensity of irrigation and preponderance of long duration varieties.

Recently one more dimension *i.e.* to maintain the soil fertility, is added in the definition of multiple cropping. Thus multiple cropping is defined as the cultivation of two or more crops on the same field in a year without deteriorating the soil fertility.

3.2 Conceptual Basis of Multiple Cropping

A cropping system is the kind and sequence of crops grown on a given area of soil over a period of time. It may be a regular rotation of different crops in which the crops follow definite order of appearance on the land or it may consist of only one crop grown year after year on the same area (monoculture). Other cropping systems may include different crops but lack definite or planned order in which crops follow one another or growing of two or several crops mixed together. Multiple cropping is thus a one year cropping system in which two (double cropping) or more crops are grown in succession within a year. If all these crops are cereals, multiple cropping will resemble monoculture in its advantages and requirements. If legumes are also included it will have certain features of traditional rotations.

A soil management system includes the cropping system with the other associated practices in raising the crops, such as frequency and method of tillage, use of fertilizers, irrigation, drainage etc. Thus effect of cropping system on soil productivity depends on other practices used with it in making up the systems of soil management.

Cropping system followed in major parts of the country are

1. Continuous cropping in which effort is made to maintain productivity. This is one year cropping system in which the main *kharif* crop alternates with *rabi* fallow or *kharif* fallow alternates with *rabi* crop
2. Rotation of two or more than two years involving occasional fallowing of legumes for fodder, seed or green manure
3. Mixed cropping of one year either during *kharif* or *rabi* or during both the seasons
4. Cropping systems with a lack of definite planned order, and
5. In certain pockets shifting cultivation in which soil is cleared, cropped for three or four years and then abandoned.

This classification is of course, too generalized and is not exhaustive. There is considerable and necessary overlap between these systems. Some of these will continue in the dry farming area but, wherever water has been made available either in dry farming area or humid areas all will give way to multiple cropping system. During the recent years a renewed emphasis has been placed on minor irrigation programmes in almost all states. All these minor irrigation sources are destined to become nucleus around which a programme of multiple cropping will develop.

For these traditional cropping systems soil management programmes were based on occasional fallowing, green manuring, frequent cultivation during fallow period to remove weeds and hot weather cultivation to speed up the burning of organic matter for releasing nitrogen. Experimentally, continuous wheat with *kharif* fallow has given steady yield of nearly 900 kg/ha. In North India, monoculture of paddy over centuries has given similar yield without manuring. Thus, wheat and paddy yields have stabilized on lower level and maintained through the operation of natural nitrogen fixing process. The year to year variation in yield around these averages is due to seasonal variation. Green manuring-wheat rotations have shown no long term effect on soil productivity. In short term they have given response equal to the application of 20-30 kg N/ha. Old rotational trials have shown the value and utility of rotations under the management followed. Long term trials at Pusa, Coimbatore, Cuttack have shown that plots receiving only balanced inorganic fertilizers have maintained yield and organic matter content of the soil. Since nitrogen was almost universally deficient for maximum yields of non-legumes there was an urgent need of including a legume in rotation and rotations like sorghum-wheat were utter failure without manuring. The whole picture has changed since nitrogen has become more and more available.

The importance of good tilth for high yield has been generally accepted but seldom found in research results to be as important as it appears. On sandy soils the tilth is seldom a problem. On medium textured soil (silt loams and loams) an acceptable level of soil tilth can be maintained in the plough layer without bringing legumes in the cropping system with minimum tillage. On fine textured soil (silty clays, clay and clayey loams) favourable tilth is more necessary than in coarse and medium textured soils. On such soils, problem of structure may become more severe when one grows the crops which leave residues to the soil. Farmers, who have such soils, should watch carefully the water retention in their soil. If seed bed is cloddy or field dries slowly after rains or water stands between the rows in low places then the soil management system used is not maintaining soil structure. More fertilizer to grow more residues or less tillage or both are needed. If paddy is included in such a cropping system to alternate with upland crop, it should be grown like upland crop without puddling. One agronomic fact has always to be remembered *i.e.* any damage from over cropping is temporary and can be corrected by changing the soil management system which provide greater return of organic residues, less frequent tillage and perhaps more deep rooted crops.

As far as the economic factors are concerned, the cost benefit ratios are more favourable now for following multiple cropping. Improved fertilizer technology, higher analysis of fertilizers to reduce handling and transportation costs, more efficient application techniques including bulk handling of dry mixture etc. are expected further to reduce cost. The peak labour requirement for harvesting of crops and sowing of *rabi* crops is generating a demand for labour saving machines. Thus the multiple cropping programmes will not only accelerate the chemicalization of the Indian agriculture, but it would also lead to mechanization. Double cropping with no or inadequate manuring or with inclusion of a legume has been practiced for long in the country. What is needed is to provide more purchased inputs, the economics is in favour.

Although written in another context the basis of multiple cropping according to Melsted (1954) is, it is a philosophy of maximum crop production per acre of land with a minimum of soil deterioration. The philosophy is based on the concept that high production is good for soil, that minimum tillage promotes soil tilth and conserves soil organic matter, high fertility promotes high yield, and lessens the losses of soil humus. Large amount of decomposable organic matter in soil is essential to good soil tilth and good soil physical conditions, and cover in the form of living mulch is good protection against all forms of erosion.

Many of the concepts on which multiple cropping is based are only tentative. The preliminary results which have been achieved in national demonstrations, by progressive farmers and in experimental plots are encouraging. They seem to indicate that the concepts are sound and future will support them. When multiple cropping will be adopted on large scale many soil management problems are likely to come up. For their solution integrated research done by plant breeders, agronomists, soil scientists and agricultural engineers will be absolutely necessary.

3.3 Pre-requisites for Multiple Cropping

A few of the conditions considered to be pre-requisite to successful multiple cropping are enumerated here under.

i. There must be adequate supply of available water and it must be available at the time it is needed.

ii. Adequate and timely supply of seed, fertilizer and plant protection materials

iii. The use of short duration varieties and hybrids. Crops of short duration are generally more useful, taking into consideration the yield per unit area per unit time rather than the longer duration ones.

iv. Development of new market, processing and new uses of the crops which are not traditionally grown in the locality but have a comparative advantage during some period.

v. High man hour requirement, availability of machinery and money

vi. Better under standing of scientific agriculture

3.4 Multiple Cropping

The intensification of cropping in time and space dimensions. It is growing two or more crops on the same field in a year. Different types of multiple cropping are

Inter Cropping

Mixed Cropping

Sequential or sequence cropping or non-overlapping Cropping.

Relay cropping or overlapping Cropping.

Double, Triple and Quadruple Cropping

Ratoon Cropping

Multi-storey / Multilevel / Multi-tier Cropping

Paira / Utera Cropping1

3.5 Inter-cropping

It refers to growing of two or more crops simultaneously, on the same piece of land, base crop necessarily in distinct row arrangement. Crop intensification is in both time and space dimensions. There is intercrop competition during all or parts of crop growth.

Or

Intercropping is the cultivation of two or more crops simultaneously on the same field, with or without a row arrangement (row intercropping or mixed cropping). Relay intercropping is the growing of two or more crops on the same field with planting of the second crop after the first one has already flowered. The crops grown in intercropping are called a crop association, and are denoted with a + sign. The crops involved can be annuals only, a combination of annual and perennials or perennials only. Important intercropping systems for various regions are given in Table 1.

Cropping intensity in space dimension is achieved. Such crops are usually grown along with those crops which have larger row interval. Intercropping includes alley cropping, strip cropping, contour cropping, paired row cropping, skip cropping, parallel cropping, companion

Table 1. Recommended intercropping systems

System	Ratio
Wheat + Raya	8 : 1 or 12 : 4
Winter maize + Peas (green pods)	1 : 1
Potato + Mustard	3: 1
Potato + Wheat + Summer mung	1 : 2: 2
Mustard + Gram	4: 1
Autumn cane + Wheat	1 : 2
Maize + Green/Black Gram	1 : 1
Maize + Soybean	1 : 1
Pigeon pea + Green/Black gram	1 : 1

cropping, multi-storeyed cropping and synergetic cropping. But the major intercropping systems are

1. *Parallel Cropping:* Cultivation of crops which have different natural habit and zero competition e.g. Black gram/green gram + maize

   The peak nutrient demand period for black gram or green gram is around 30-35 days after sowing while it is 50 days after sowing for maize.

2. *Companion Cropping:* Such intercropping where the production of both intercrops is equal to that of its solid planting. e.g. Mustard/potato/onion + sugarcane.

   By growing of the above three intercrops individually with sugarcane, the production of none is hardly affected.

3. *Synergetic Cropping:* The yields of both crops are higher than of their pure crop on unit area basis e.g. sugarcane + potato. On the basis of per cent of plant population used for each crop in intercropping system; intercropping is of two types *viz,* additive series and replacement series.

a. *Additive series intercropping:* In such type of intercropping, one crop is main crop or base crop and another crop is intercrop. Intercrop is introduced into the base crop by adjusting or changing crop geometry *i.e.* addition of intercrop to the base crop. Here plant population of base crop is same to what recommended population in pure stand whereas, that of intercrop is less. Cropping husbandry is according to the base crop. This type of intercropping is prevalent in India and its main objective is to get additional income and to cover risk. e.g. sowing of potato in the field of sugarcane in between the rows of cane at 90 cm.

b. *Replacement series intercropping:* Both the crops are component crops. Neither is the base crop nor is the intercrop. It means the plant population of both component crops is less than their recommended population in pure stand. Such type is widely practiced in western countries. Here 10th row of wheat crop is replaced by mustard and row to row distance of wheat crop is somewhat reduced to adjust the mustard crop.

Row intercropping

Growing two or more crops simultaneously, where one or more crops are planted in rows. Often simply referred as intercropping.

Patch cropping

The component crops are planted in patches

Strip intercropping

Growing two or more crops simultaneously in strips wide enough to permit independent cultivation but narrow enough for the crops to interact agronomically.

Main crop / Base crop

This is the one which is planted at its optimum sole crop population in an intercrop situation and the second crop is planted in between rows of the main or base crop with a view to obtain some extra intercrop yield without sacrificing the main or base crop yield.

Component crop

The component crop is used to refer to either of the individual crops making up the inter-cropping situation.

3.6 Pre-requisites of Intercropping

The main objective of intercropping is higher productivity per unit area in addition to stability in production. Intercropping utilizes resources efficiently. There are certain important pre-requisites of intercropping,

- The time of peak nutrient demands of component crops should not overlap
- Competition for light should be minimum among the component crop
- Complimentarity should exist between the component crop
- The differences in maturity of component crops should be at least 30 days
- Competition for CO_2 and water should be minimum among component crops

3.7 Mixed cropping

Cultivation of two or more than two crops simultaneously on the same land without definite row pattern or fixed ratio. Sowing of seeds is generally by broadcasting method commonly practiced in dryland area of India. It is subsistence farming. Main objective is to lessen the risk of total crop failures, and to satisfy the farmers in

Advantage and disadvantage of intercropping

Advantages	Disadvantages
Better use of growth resources including light, nutrients and water. Suppression of weeds	Labour intensive
Reduced pest and disease incidence	Control of pests and diseases or chemical weed control may be difficult
Yield stability	Mechanization is difficult
Ecological stability *i.e.* improvement of soil health and agro-ecosystem E.g. Spreading of labour, physical support of one crop to another and home gardening leading to a more food supply	Disadvantages caused by adverse competitive effects or by allelopathy

food and fodder. Intercropping is the new concept of mixed cropping. Scientific study of mixed cropping was firstly done by La-Flitze (1928).

3.8 Sequence/sequential Cropping or Non-overlapping Cropping

Growing of two or more crops in quick succession, on the same piece of land in a farming year. The sowing of the succeeding crop and harvesting of the preceding crop may be done simultaneously or in a quick succession. e.g. just after the harvesting of maize, potato is sown and just after digging of potato, chilli is sown. Sequential cropping is also called non-over lapping cropping because there is no overlap between the two or more cropping. Ratooning or ratoon cropping may also be classified under sequential cropping.

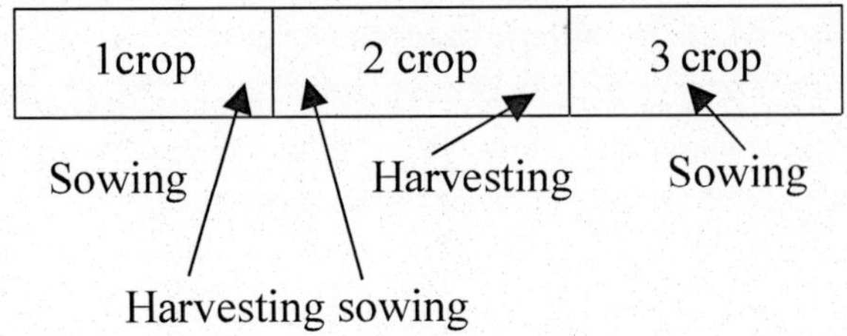

3.9 Relay Cropping or Overlapping Cropping

This concept has been derived from relay race in which four runners run in the field having flag in their hand. The first runner passes on his flag to the succeeding partner and 2nd to 3rd and 3rd to 4th runner. Relay cropping can be defined as growing two or more crops simultaneously during the part of the life cycle of each. Succeeding crops are planted before harvesting the preceding crop. Generally, 2nd crop is planted after the first crop has reached its reproductive stage of growth but before it is ready for harvest e.g. Potato is planted before the harvesting of Maize and Radish is sown before harvesting of potato. Paira (Bihar & W.B.) or Utera (M.P) cropping may be example of relay cropping. Paira or utera cropping means sowing of lathyrus or lentil (generally pulses) before the harvesting of rice in lowland area. Where the land after harvest of paddy remains slaucy, making it unsuitable for tillage, these crops may be broadcasted atleast 15-20 days before harvest using 1.50 times the seeds required for normal sowing. Planting of nursery crop may also be an example. Sher Singh Bains has suggested the following relay cropping pattern for North-West part of India.

Mung - Maize - Potato - Wheat

I crop	II crop		III Crop		IV Crop	
Sowing of 1 crop	Sowing of 2 crop	Harvesting of 1 crop	Sowing of 3 crop	Harvestin g of 2 crop	Sowing of 4 crop	Harvesting of 4 crop

Double, Triple and Quadruple Cropping

Double cropping Growing two crops in sequence in a year.

Triple cropping : Growing three crops a year in sequence

Quadruple cropping: Growing four crops a year in sequence.

Ratoon cropping

The cultivation of crop re-growth after harvest, although not necessarily for grain.

3.10 Multi Storey / Multilevel / Multi-tier

Cultivation of two or more than two crops of different heights simultaneously on a certain piece of land in any certain period. e.g. Sugarcane + Mustard + Onion / Potato. Intercropping is mostly prevalent in plantation crops like coconut and arecanut. There is scope for intercropping in coconut garden upto the age of 8 years and after 25 years. During this period, there is adequate light transmission to the ground which permits cultivation of intercrops. The practice of growing different crops of varying height, rooting pattern and duration is called 'multi-tier cropping'. The objective of this system of cropping is to utilize the vertical space more effectively. In this system, the leaf canopies of intercrop components occupy different vertical layers. The tallest components have foliage tolerant of strong light and high evaporative demand and the shorter component (s) with foliage requiring shade and or relatively high humidity.

Eg., Coconut + black pepper + cocoa -+ pineapple. The schematic representation of planting of these crops is shown in Fig.

In this system, coconut is planted with a spacing of 7.5 m. Rooted cuttings of black pepper are planted on either side of coconut about 75 cm away from the base. On the

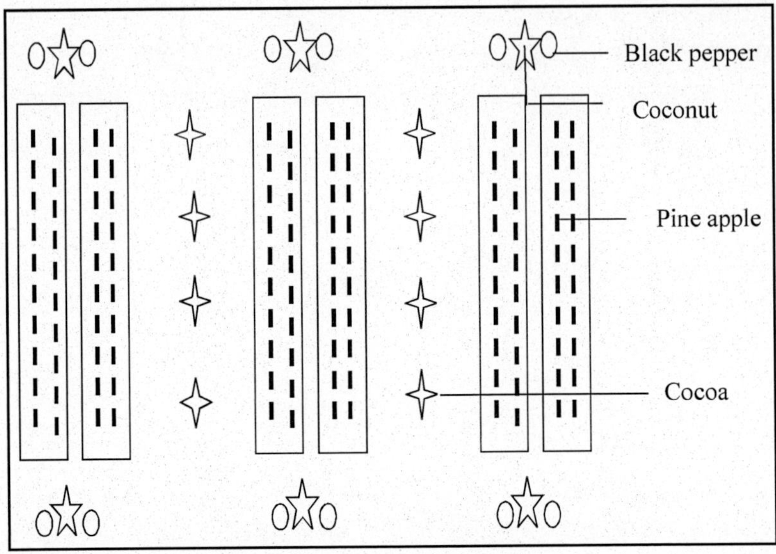

Fig. Multi storeyed cropping

coconut trunk at a height of about one metre from the ground level, the vines of pepper are trailed. A single row of cacao is planted at the centre of space between coconut rows. Pine apple is planted in the interspace

Coconut growing to a height of more than 10 m occupies the top floor. Black pepper is growing to about 6-8 m height which forms the second floor.

Cocoa with its pruned canopy of about 2.5 m height and pine apple growing to about 1 m height form the first and ground floors respectively.

In another multi-tier system in coconut, ginger or turmeric and partial shade loving vegetables form the first tier, banana in second tier, pepper in third tier and coconut or arecanut in the final tier.

In the arecanut plantation, tuber crops are predominantly intercropped. Elephant yam, tapioca, greater yam and sweet potato are common intercrops in humid tropics. Banana and pine apple are also cultivated as intercrops in arecanut gardens.

In coffee based multi-tier cropping system, first tier is with pine apple, second tier with coffee, third tier with cacao/mandarin orange and final tier with fast growing shade trees necessary for coffee plantation.

3.11 Choice of Crops for Multiple Intensive Cropping

1. The crops should be arranged in such a way that allelopathic effect, temporary immobilization of nutrients and depletion of nutrients from the same layer of soil do not occur.
2. Legumes have a definite place in any cropping system because of their limited demand on resources like water, nutrients and light, their adaptability to varying environmental conditions and capacity to fix atmospheric N in root nodules.
3. Vegetables deserve their due place in intensive cropping because

   a. of high cash and nutritional values

   b. most vegetables can be raised as seedlings and planted

   c. reduces main field duration and competition with the associated crop

   d. they can be grown in interspaces in orchards and plantations

   e. they can be inter or relay cropped in cereal based crops
4. Short duration, photoperiod insensitive genotypes are suited for intensive cropping system

4. CROP ROTATION

Crop rotation is a process of growing different crops in succession on a piece of land in a specific period of time, with an object to get maximum profit from least investment without impairing the soil fertility. One cycle may take one or more farming years to complete. Farming year is 12 months for irrigated areas and is limited to the period of adequate soil-water availability for crop growth in rainfed areas.

4.1 Principles of Crop Rotation

- The crops with tap roots should be followed by those which have a fibrous root system. This helps in proper and uniform use of nutrients from the soil.
- The leguminous crops should be grown after non-leguminous crops because legumes fix atmospheric N into the soil and add more organic matter to the soil.
- More exhaustive crops should be followed by less exhaustive crops because crops like potato, sugarcane, maize etc. need more inputs such as better tillage, more fertilizers, greater number of irrigations etc.
- Selection of the crop should be demand based.
- The crop of the same family should not be grown in succession because they act like alternate hosts for insects, pests and diseases.
- An ideal crop rotation is one which provides maximum employment to the family, the machines and equipments are efficiently used and all the agricultural operations are done timely, simultaneously maintaining soil productivity.
- The selection of the crops should be problem based
- On sloppy lands which are prone to erosion an alternate cropping of erosion promoting and erosion resisting crops like legumes should be adopted.

- Under dry faming the selection of crops should be such that, crops can tolerate drought.
- In low-lying and flood prone areas the crops should be such that which can tolerate water stagnation. .
- The selection of crops should suit farmer's financial conditions. The crop selected should also suit the soil and climatic conditions.

4.2 Advantages of Crop Rotation

- Agricultural operations can be done timely for all the crops because of less competition.
- Soil fertility is restored by fixing atmospheric nitrogen, encountering microbial activity, avoiding accumulation of toxins (HCN etc.) and maintaining physical properties of soil.
- An ideal crop rotation helps in controlling insects, pests and diseases. It also controls the weeds in the fields.
- Proper utilization of all the resources and inputs could be made by following crop rotation.
- The farmer gets a better price for his produce because of its higher demands in the locality or market.

Crop rotation is the reverse of land rotation. Here land is fixed but crop is rotated year after year. On a certain land, repeated cultivation of crop or crop fallow land in a certain sequence is called crop rotation. Eg., Rice - Wheat - Mung (Green gram), Rice - Mustard - Chilli. Maintains and even improves soil fertility and stabilizes income.

It checks the soil erosion and can serves moisture. Ensures a balanced programme of work throughout the year. Suitable crop rotation is the key of modern scientific agriculture which aims to produce maximum yield by maintaining soil productivity.

5. PLANT INTERACTIONS IN CROPPING SYSTEM

In intensive cropping, crops are grown in association (intercropping) or in sequence (sequential cropping). In such situations there is possibility of interaction between the component crops. The interaction is mainly due to response of one species to the environment as modified by the presence of another species. Interaction may be competitive or non-competitive or complementary. If the crops are grown in association and the growth of either of the concerned species is not affected, such type of interaction is called as non-competitive interaction or interference. If one species is able to help the other it is known as complementary interaction. e.g. supply of N fixed by the legumes to the associated non-legumes. (This is otherwise known as annidation). One species may have greater ability to use the limiting factor and will gain at the expense of the other. This is called as competitive interaction or interference. Interaction may also occur in some other manner by way of producing toxic substances and affecting the establishment and growth of the associated species. This is called allelopathy.

5.1 Interactions in Mixed Crops

Plants require light, water, nutrients, oxygen and CO_2 for growth. Light and CO_2 are absorbed by the leaves, and water, nutrients and oxygen mainly by the roots. Growth is responsive to the rate of absorption of any one of them when it is in relatively short supply.

In a young sole crop, the leaf area index (LAI) and root density are low and each plant is able to get its requirement as through every one of them is growing in isolation. As the plants grow, LAI and root density increase leading to mutual interference in the absorption of one or more growth factors. Growth rate per plant falls below that obtained from plants grown in isolation (without interference). The close proximity of neighbours causes sub-optimal absorption of growth factors and so there is inequitable distribution of resources among the plants. The reduction in per plant growth and yield is compensated by the overall production from a large number of plants growing in a unit area. Hence, at the optimum plant density of sole crops, the production per plant is sub-optimal but the production per unit area is optimal.

Introduction of another plant species by replacing part of the population of one species (replacement series) as in 1:1 or 2:1 ratio planting or by adding the second species without reducing the population of the first species from the optimum (additive series) causes complex interferences between the species in addition to the competition described above. Early interference among plants is similar both in sole and in mixed crop communities since the plants are sufficiently far apart. Gradually competition will develop between the two components and hence "between component competition" will be added to "within component competition". The nature of such competitions will vary depending on the density and proportion of each component and planting pattern. The "between component competition" will be affected by the spatial and temporal variations between the components.

5.2 Interactions in Intercropping

Factors such as light, water, nutrient, oxygen and CO_2 are required for plant growth. In mixed or intercropping situations, the component species compete for the growth factors. The close proximity of species causes sub-optimal utilization of the growth factors and hence there is inequitable distribution of resources among the plants. Generally competition will develop between two components or within the components. The nature of such competition may vary depending on the density and proportion of each component and planting pattern.

5.2.1 Light

Photo-synthetically active radiation (light) is directly related to photosynthetic rates of component crops. Intercropping can increase light interception by as much as 30-40%. When one component is taller than the "other in an intercropping system, the taller components intercept most of the solar radiation. The pattern of light interception also varies according to the age of the crop. Light transmission by coconut is only about 20% at the age of 8-10 years and this remains almost constant till about 25 years of age. Subsequently the % light transmission increases progressively as the canopy coverage of ground decreases. When the palms are at about 40 years of age, light transmission increases to about 50%. In an intercropping situation where the component crops have different growth durations, the peak demand for light would occur at different times. For example, in maize + green gram intercropping, the peak light demand for maize occurs at 60 days when green gram is ready for harvest. In such combinations, competition for light is less among the component crops and there is greater light use in intercropping than in pure stands. In general, the component crops under intercropping situations are grown in such a way that competition for light is minimized. Proper choice of crops and varieties, adjustment of planting density and pattern are the techniques to reduce competition and increase the light use efficiency.

5.2.2 Moisture and nutrients

Competition for moisture and nutrients may result in two types of effects on the less successful or suppressed component. First, the roots to this component may grow less on the sides towards plants of aggressive component. Secondly, plants affected by competition for soil factors may have increased root/shoot ratio. The aggressive component generally absorbs greater quantity of nutrients and soil moisture. In legume and non-legume combination, the later takes up large amount of P, K and S. As a result, the legume may show deficiency of these nutrients. Such effects, however, may be mitigated by appropriate fertilizer application. Among intercrops, sorghum and pearl millet are more competitive in extracting nutrients. Generally the intercropped stand remove greater amount of nutrients than sole crop stands.

5.2.3 Allelopathy

Allelopathy is any direct or indirect harmful effect that one plant has on another through the production of chemical substances that escape into the environment. Some crops may be unsuitable to be grown as intercrops because they may produce and excrete toxins into the soil which are harmful to the associated components. Allelochemicals produced from the leaves of *Eucalyptus globules* drastically reduce the germination of mustard sown underneath.

5.2.4 Competition for solar radiation

Solar radiation falling between the wavelengths of 300 and 700 nm is referred to as photosynthetically active radiation (PAR) and that falling above 700 nm is called the near infrared radiation (NIR). PAR is commonly known as light and is directly related to photosynthetic rates of component crops while NIR is the primary energy source driving evapo-transpiration, sensible heat exchange and other photomorphogenic processes.

5.2.5 Competition for carbon dioxide

Although competition between component crops for CO_2 is theoretically possible, the turbulence within canopies is usually so great that it seems unlikely to occur. So far it has been demonstrated only within sealed enclosures

5.2.6 Competition for soil factors

Water, nutrients and oxygen are mainly taken up by the roots from the soil and so as considered together as soil factors. In the early stages for growth, roots will be far apart for uptake by one radicle not to interfere with the supply of soil factors to its nearest neighbour. However, since the surface area of the root system is larger than that of the shoot, the soil soon becomes crowded and competition for supplies begins.

5.2.7 Annidation

Annidation refers to complementary interaction which occurs both in space and time.

Annidation in space: The canopies of component crops may occupy different vertical layers with taller component tolerant to strong light and high evaporative demand and the shorter component favouring shade and high relative humidity Multi-storeyed cropping in coconut and planting shade trees in cocoa and tea plantations use this principle. Similarly, root systems of component crops may exploit the nutrients in different layers of soil and hence utilize the resources in a better way with much less competition.

Annidation in time: When component crops of widely varying duration are planted, their peak demand for light and nutrients are likely to occur at different periods, thus reducing competition. In a combination having early and late maturing crops (sorghum + red gram), when early maturing crops are harvested, conditions become favourable for the late maturing crop (red gram) to put forth its full vigour.

5.2.8 Other complementary effects

In an intercropping system involving a legume and a non-legume, part of the N fixed in the root nodule of the legume may become available to the non-legume component. With the presence of rhizosphere microflora and mycorrhiza, one species may lead to mobilization and greater availability of nutrients not only to the species concerned but also to the associated species. Provision of physical support by one species to the intercropped climbing species may improve the yield of the climber. Examples are coconut + pepper and maize + beans. The taller component acts as wind barrier protecting the shorter components from lodging.

Multiple cropping in the form of intercropping is predominant in the regions of dry, humid and semi-arid tropics. Farmers normally grow two or more crops together on the same piece of land. Different crops are cultivated in mixed cropping. Simple mixture of only two or more crops is the most common.

The objectives of this practice are:

1. Insurance against total crop failure under aberrant weather conditions or pest epidemics.
2. Increase in total productivity per unit land area.
3. Judicious utilization of resources such as land, labour and inputs.

When two crops are to be grown together, they are chosen in such a way that there is variation in their growth duration. The peak periods of growth of the two crop species should not coincide. In such arrangements, a quick maturing crop completes its life cycle before the other crop starts. Willey (1979) described the concept as temporal complimentarity. Greater the differences in maturity and growth demands of the crop components, more opportunity is provided for greater exploitation of growth factors and over yielding. This will be achieved either by genetic difference in crop species or manipulation of planting dates. Normally short and long duration crops are grown together.

Food crops are usually mixed with cash crops to ensure both sustenance and cash income.

Combination of species under intercropping depends on climate, location preferences and other site specific factors. Some of the important intercropping systems are described below:

5.2.9 Intercropping in cereals

Intercropping with cereal is an excellent way of improving the resource utilization because the cereal utilizes the rainy season resources while late maturing crop exploits the post-rainy season resources such as residual moisture. Sorghum is most commonly intercropped with pigeon pea on a variety of soils. Sorghum is harvested after 105 days to 135 days and pigeon pea matures in about 6-9 months depending on the genotype. Grain legumes such as cowpea, green gram, black gram and groundnut are the other recommended intercrops with sorghum.

Maize + bean *(Phaseolus vulgaris* L.) is a predominant intercropping system. In Central and South America and parts of East Africa, where rainfall ranges from 700 to 1500 mm spreading over 4-5 months. Soybean is also a good compatible companion crop with maize.

Pearl millet is a quick tillering and fast growing crop that attains full canopy development within 20-30 days of seedling establishment. It can be intercropped with groundnut, black gram or castor.

Finger millet is widely intercropped with field beans *(Dolichos lab- lab)* in Kamataka State, India, where 4 - 12 rows of the millet are alternated with one row of beans.

5.2.10 Intercropping in pulses

For pigeon pea, short duration grain legumes such as black gram and soybean are the best companion crops in

Peninsular India. Groundnut is also a suitable intercrop. Sorghum is commonly intercropped with pigeon pea.

5.2.11 Intercropping in cotton

Cotton is initially a slow growing crop. Any short duration and fast growing crops such as green gram, black gram, groundnut or cluster bean are the compatible companion crops. Onion is also a preferred intercrop in cotton.

5.2.12 Intercropping in sugarcane

Sugarcane is slow growing upto 80 - 90 days. Since the crop is planted in rows in 0.8 - 1.0 m apart, considerable space is available for intercropping. Short duration crops maturing in 80-90 days can be advocated as intercrops. Black gram and soybean are found suitable. The green manure, daincha *(Sesbania aculeata)* can be sown in the interspace and incorporated at about two months of its growth.

5.2.13 Intercropping for drylands

Intercropping recommended for different regions of India under drylands conditions are given in Table 2.

5.3 Sequential Cropping System

In sequential cropping, sole crops are grown in succession. Hence, competition for light, water and nutrients as found in mixed cropping does not occur. In relay cropping, a short span of overlapping occurs between two crops in sequence. Hence, the relay sown crop at the seedling stage may have competition for light from the standing crop. Such type of competition may be minimized by proper choice of crops and varieties and adjustment of time and method of planting.

In intensive multiple cropping involving two or more crops in sequence, the main objective is to harvest as much

Table 2. Intercropping for different regions in India

Region	Intercropping
Samba (Jammu)	Castor + black gram / sesame / cowpea / sunflower / Maize + cowpea (fodder) Maize + green gram
Hissar (Haryana)	Pearl millet + cluster been / green gram
Dehra Dun (U.P.)	Pigeon pea + soybean Maize + pigeon pea
Agra (UP.)	Pearl millet + black gram / green gram Pigeon pea + green gram / pearl millet
Varanasi (U.P.)	Pigeon pea + maize / black gram / millet
Ranch (Bihar)	Barley + rapeseed Wheat + Bengal gram / rapeseed / lentil
Bhubarieswar (Orissa)	Pigeon pea + black gram / sorghum / groundnut
Rewa region (M.P.)	Pigeon pea + sorghum + finger millet / black gram / sunflower
Jhansi (M.P.)	Sorghum + black gram / green gram / soybean Wheat + Bengal gram / mustard Barley + Bengal gram / mustard Bengal gram + linseed / mustard
Indore (M.P.)	Pigeon pea + sorghum Cotton + soybean

Contd...

Table 2. Contd...

Udaipur (Rajasthan)	Sorghum + black gram Pearl millet + black gram Sunflower + black gram
Anand (Gujarat)	Pearl millet + red gram / castor / sunflower
Akola (Maharashtra)	Pigeon pea + black gram / green gram / sorghum / groundnut Cotton + sunflower / black gram / groundnut
Sholapur (Maharashtra)	Pigeon pea + pearl millet / sorghum / black gram Cotton + sunflower / groundnut
Anantapur (A.P.)	Pigeon pea + green gram / pearl millet / sorghum / foxtail millet / groundnut / sunflower
Bijapur (A.P)	Cotton + groundnut / sorghum / pearl millet / green gram
Mysore region (Karnataka)	Pigeon pea + pearl millet, Cotton + pearl millet Groundnut + cotton
Kovilpatti (T.N.)	Cotton + black gram / green gram Sorghum + black gram / cowpea / *Dolichos lablab*, Pearl millet + green gram / cluster bean / cowpea / *Dolichos lablab*

solar energy per unit area per unit time as possible. In rice based cropping system, the solar energy use efficiency ranged from 1.58 to 2.02% of PAR in Uttar Pradesh. In Coimbatore, the efficiency ranged from 0.82 to 1.77% of PAR for the whole year in cotton based system. Inclusion of a C4 plant in the summer increased the efficiency.

In sequential cropping, the preceding crop has considerable influence on the succeeding crop. This includes the complementary effects such as release of N from the residues of the previous crop, particularly legume, to the following crop and carry over effects of fertilizer applied to the preceding crops. The adverse effects include allelopathy, temporary immobilization of N due to wide C/N ratio of the residues and carry over effect of pest and diseases. Introduction of pulses as summer crops results in improved yield of rice in *kharif* season.

In India, food crop is predominantly grown in most suitable seasons and thus particular food crop is basic to the cropping system followed by the farmers. Accordingly the cropping systems are usually referred to as

a. Rice based cropping system
b. Sorghum based cropping system
c. Pearl millet based cropping system
d. Wheat and gram based cropping system

Some of the cropping systems based on commercial crops are (i) cotton based, (ii) groundnut based (iii) sugarcane based, (iv) plantation crop based, and (v) vegetable based cropping system.

5.3.1 Cropping system under wetlands

Sequential cropping is the most suitable cropping pattern of rice-rice in the rice areas. Traditionally single crop of rice with long duration cultivars is planted during the rainy period. However, due to the advent of short duration, photo-insensitive varieties, multiple cropping is being now practiced. The cropping system adopted under lowland condition can be grouped into two broad categories such as all rice and mixed rice and upland crop sequential cropping systems. These can be further grouped as below.

All rice cropping pattern

1. Rice- Rice (R-R)
2. Rice- Rice - Rice (R -R -R)

Mixed rice and upland crop sequence

1. Rice -upland crop (R-UC)
2. Upland crop - Rice -upland crop (UC-R-UC)

All rice sequential cropping system

Rice crop is repeatedly grown wherein two or three crops are raised. With three crops of rice, the land is cropped at least ten months in year. This is possible only in irrigated areas where water is adequate even during the dry periods. Crop failure due to water stress normally does not exist. Among the three crops, productive rice crop is the one which coincides with the dry months. During this period, solar radiation is high and the occurrence of calamities such as cyclone or pest infestation is very low.

Two rice crops are mostly cultivated in many parts of the rice growing areas in India. The first crop is a short duration crop of rice which is cultivated during the *Kharif* season. This is followed by the second crop of rice during *rabi* season with medium duration varieties. Single crop of rice with the medium or long duration varieties is most prevalent in the command areas. Cultivation is restricted to the period of water avail1ability in the canal. All rice sequence is followed in some pockets of the river command areas of Orissa, Andhra Pradesh, Karnataka and Tamil Nadu.

Promising rice-based cropping systems in India

Promising rice-based cropping systems with high productivity and intensive land use for diverse agro-climatic situations of India have been identified in multi-locational experiments under Cropping Systems Research project (Table 3).

Table 3. Promising rice based cropping systems in India.

System	Region
Rice - Indian mustard	Kashmir valley
Rice - wheat - sorghum (Fodder)	Jammu region
Rice - wheat - green gram	Punjab, Haryana, U.P.
Rice – Mustard	North Konkan
Rice - field bean	South Gujarat
Rice – groundnut	South Karnataka
Rice - black gram/sesame/cotton	Thanjavur, Tamil Nadu (Old Cauvery delta)
Rice – groundnut	Thanjavur, Tamil Nadu (New Cauvery delta)
Rice – mustard	Assam
Rice - wheat – jute	West Bengal
Rice - ground nut	Coastal Andhra Pradesh and Orissa
Rice - green gram	Eastern Madhya Pradesh

Systems with more than three crops

With ratooning or relay cropping and special cultural practices that avoid puddling in lowland rice, upto five crops can be harvested in the same rice field over 12 month period (Bradfield, 1970). In rice fields with adequate irrigation and good water control, Bradfield experimented at IRRI, Philippines, with various cropping patterns that could maximize productivity per unit area. The most successful pattern was a sequence of four crops consisting of rice, sweet potato, soybean and sweet corn. The tillage pattern varied for each crop and number of days to harvest were 102, 100, 85 and 66 respectively. This pattern gave three times higher gross income than normally obtained in a rice- rice pattern.

5.3.2 Cropping system under irrigated uplands

Under irrigated uplands (garden lands) depending upon the water availability or irrigation potential, more than two crops can be grown in a year. Crop sequences and their production potential have been tested under different agro climatic conditions. At Delhi, four crop sequence of green gram -maize - potato- wheat gave the highest production of 13.6 t/ha. However, three crop sequence of green gram - maize - wheat was found to be the most ideal for the small and marginal farmers (Table 4).

Table 4. Crop sequences and their production potential, New Delhi, India

Crop sequences	Cropping pattern	Total production (q/ha)
Two crops	Maize – wheat	90.1
Three crops	Green gram - maize - wheat	97.7
	Green gram - pigeon pea - wheat	68.8
	Cowpea - pearl millet –wheat	92.9
Four crops	Green gram - maize - potato -wheat	135.9
	Green gram - maize - toria -wheat	112.8

Under All India Co-ordinated Agronomic Experiments Project, Production potentials of the crop sequences under irrigated condition were studied. Some of the important sequences and their yield potential are given in Table 5.

In multi - location experiments, under Cropping Systems Research Project, promising cropping systems with high productivity as well as intensive land use for diverse agro climatic situations have been identified (Table 6).

Table 5. Crop sequences in different states of India

State	Crop sequences
Himachal Pradesh	Soybean – wheat
	Soybean – potato
	Maize – wheat
Haryana	Pearl millet
	wheat - cowpea (f)
	Maize – wheat
	cluster bean (f)
	Pearl millet
	wheat - maize (f)
Uttar Pradesh	Maize – wheat
	Sorghum – wheat
	green gram
	Sorghum – wheat
	Cowpea
Rajasthan	Pearl millet
	wheat - green gram
	Green gram
	wheat - green gram
Madhya Pradesh	Maize – wheat
	Cowpea
	Maize – wheat
	Green gram (f) - fodder

5.3.3 Cropping system in drylands

Normally only one crop is grown under dryland condition and cultivation is restricted during the rainy

Table 6. Potential cropping systems for different agro climatic situations

Region	System
Western Haryana	Cotton – wheat
Rajasthan	Groundnut - wheat
Madhya Pradesh	Soybean - wheat
North Gujarat	Pearl millet - mustard
Sourashtra region - Gujarat	Cotton - groundnut
Western Maharashtra	Sorghum - wheat
Western Maharashtra	Groundnut - wheat
Marathwada region of Vidarba region	Cotton - groundnut
North Karnataka	Sorghum - bengalgram
South Karnataka	Cotton - groundnut

season. However, the intensity of cropping can be increased through sequential or intercropping, depending on the rainfall and moisture storage capacity of the soils (Table 7).

Table 7: Cropping pattern with varying rainfall and soil moisture storage capacity

Rainfall (mm)	Storage capacity of soil (mm)	Cropping pattern
350 – 625	100	Single crop in *Kharif*
650 – 750	100	Intercropping can be attempted
780 – 900	150	Sequential cropping is possible
900 - above	200	Sequential cropping is assured

Source: Sharma, 1980

Choice of crops: Selection of a suitable crop and genotype is important to get higher yield in dry farming conditions. Deep rooted crops extract moisture from the deeper layers and they also have slow rate of transpiration. Leguminous crops are very well adopted for rainfed conditions because of their root system. Among the cereal, barley has better ability to harvest moisture from deeper layers than wheat. Moisture loss from the plant surface is less in barley. Similarly linseed has greater ability to trap soil moisture under water scarce situation.

Important factors such as total rainfall, soil type and water availability have to be considered while practicing a particular cropping pattern in a place / region. Cropping system for different regions as suggested from the results of AICRP (1970) are given in Table 8.

Table 8. Cropping system for different region of India under dry lands

Regions / soil type	Crops	
	Kharif	*Rabi*
Northern Region		
Samba (Jammu)	Maize	Wheat
	Green gram	Barley
	Cowpea	Wheat
	Sunflower	Wheat / barley
	Groundnut	Barley
	Pearl millet	Bengal gram
Punjab	Sorghum	Wheat
	Maize	Wheat / Potato / Barley
Hisar (Haryana)	Pear millet	Cluster bean / Chickpea
Arid soils (Sierozemic soils)	(Mono-cropping is the rule)	
Central Region		
Dehra Dun (Uttar Pradesh)	Maize	Wheat
	Rice (upland)	Wheat/Bengal gram
Agra (Uttar Pradesh)	Gram	Mustard

Contd...

Table 8. Contd...

Entisols (Alluvial soils)	Pearl millet	Chickpea / Barley
Varanasi (Uttar Pradesh)	Sesame	Chickpea / Mustard
Entisols (Alluvial soils)	Bengal gram	Barley / Mustard
	Rice	Chickpea / Mustard
	Pearl millet	Chickpea / Mustard
Eastern Region		
Ranchi (Bihar)	Rice	Chickpea / Linseed / Barley
Alfisols and related red soils	Maize	Rapeseed / Bengal gram
Bhubaneswar (Orissa)	Rice	Linseed / Mustard
Alfisols and related red soils	Finger millet	Horse gram
	Maize	Horse gram
Rewa region (M.P.)	Rice	Chickpea / Lentil
Vertisols and related black soils	Sorghum	Bengal gram / wheat
	Green gram	Wheat / Chickpea
Jhansi region (M.P.)	Sorghum	Bengal gram
Vertisols and related black soils	Pearl millet	Bengal gram
	Black gram	Safflower / Chickpea
	Sesame	Bengal gram
Indore (M.P.)	Maize	Chickpea
Vertisols and related black soils	Soybean	Safflower
	Soybean	Wheat
Western Region		
Udaipur (Rajasthan)	Green gram	Safflower
(Vertisols and related black soils)	Sorghum	Bengal gram
	Pearl mille	Bengal gram
	Maize	Wheat
	Black gram	Wheat
	Sunflower	-
Anand (Gujarat)	Pearl millet	Wheat
	Pigeon pea	
	Castor	-
	Sunflower	-

Contd...

Table 8. Contd...

Akola (Maharashtra)	Green gram	Safflower
Vertisols and related black soil	Sorghum	Safflower
Sholapur / (Maharashtra) Vertisols	Green gram	Safflower
	Pearl millet	Bengal gram
	Black gram	Sorghum
Southern regions		
Anantapur (AP)	Sorghum	Safflower / Horse gram
	Pearl millet	Cowpea / Black gram
	Black gram	Sorghum
	Green gram	Pearl millet
	Cowpea	Pearl millet / Sorghum
Hyderabad	Pearl millet	Cowpea
Alfisol	Sorghum	Horse gram
Bijapur	Sorghum	Greengram / Blackgram
Vertisols	Green gram	Sorghum / Safflower
Bangalore (Karnataka)	Cowpea	Finger millet
Alfisols	Cowpea	Chilli
Mysore (Karnataka)	Groundnut	Sorghum
	Setaria	Safflower
	Pearl millet	Cotton
Bellary (Karnataka)	Cotton	Sorghum
	Black gram	Bengal gram
	-	Safflower
Kovilpatti (TN)	-	Cotton, Sorghum, Pearlmillet, Cowpea, Horse gram, Chilli

6. AGRONOMIC REQUIREMENT OF CROPPING SYSTEM

Cropping system implies an association of different crops and their interaction. An interaction is essentially a response of one species to the environment as modified by the presence of another species. Such interactions are the characteristics of any system and they greatly influence the management of the system. The interaction in sequential cropping differs quantitatively and qualitatively from those of intercropping systems.

In sequential cropping at any particular time the interaction is between crops of the same species only. Whereas, in intercropping the interactions are between the crops of the same species as well as between crops of different species. The nature of such interactions depends on the types of component crops and their proportions.

The possible effects of such interactions can be the following

Difference in spatial and time dimensions in the crop associations.

Complementary effects of crop associations

Competition for available resources like light, water and nutrients

Effect on soil fertility and nutrient availability and uptake

Possible changes in pests, disease and weeds incidence

Adverse effects due to allelopathy

Management of cropping systems must aim at increasing the complementary interactions and minimizing the competitive interactions between the component crops so that the overall productivity of the system does not suffer.

Cropping system management has to be studied under the following aspects or situations.

Selection of crops and varieties

Tillage and land shaping

Plant populations, crop geometry, method and time of planting

Water management

Weed management

Soil fertility management and fertilizer use

Plant protection measures

Labour management

Other management practices for specific objectives of cropping system

6.1 Selection of Crops and Varieties

I. Selection of crops and varieties for sequential cropping

i) Selection of crops

The nature of crops that would fit into a crop sequence at any given farm depends on a number of factors.

a. Irrigation water availability

If supplemental irrigation facilities are available in rice fallows crops like cotton, groundnut and gingelly can be grown. If it is not available only rice fallow pulse like blackgram can be raised with residual moisture.

Similarly, in garden lands, if adequate irrigation facilities available during summer, a three crop sequence of ragi-cotton-sorghum, accommodating sorghum during summer can be followed. If water is not adequate only double crop sequence of ragi-cotton-fallow is adopted.

b. Soil type

In wetlands, to increase the cropping intensity, rice fallow cropping is recommended. The type of crop to be included is decided by the soil type also. In old delta area with heavy clay soil, cotton/pulse is suitable but not groundnut. Whereas, in new delta area where the soil is sandy with light texture, crops like groundnut and gingelly can be grown.

c. Total returns

Depending upon the price and marketability as well as proximity to the town, the selection of crops will vary. In farms nearer to town, high intensity cropping is possible by including short duration vegetables, greens and fodder crops which are quickly perishable. Such crops cannot be readily included in farms located far away from the city limits. Similarly, farmers away from town grow crops which can be stored for a longer period and upto receiving fairly good prices.

d. Multiple requirements of the farm and farmer

The various crop sequence decided for a farm should include crops which provide cereals, millets, pulses, cash crops of longer and shorter duration and also fodder crops. An ideal cropping scheme in a farm must have about 60-70% area under commercial crops 30% food crops and 10% with fodder crops.

e. Maintain soil fertility

The nature of crops selected in the intensive cropping should help to maintain the soil fertility and not exhausting it completely. There should be at least one legume crop in the cropping to maintain the fertility of soil. If a particular crop is grown season after season repeatedly, same amount of nutrients is removed from the same layer of soil. Therefore, inclusion of a legume in crop sequence at least for a short period is desirable or crops to fallow should differ from the previous crop in rooting pattern, types of nutrients required, etc., so that the soil layer or some nutrient is not continuously exploited leading to poor soil fertility.

f. Crops

Crops whose *decomposing residues* are likely to cause *toxic effect* on the germination and establishment of the succeeding crops must be avoided. If included, sufficient time for their full decomposition or proper residue management must be made.

Eg. Sunflower stalks inhibited the establishment and growth of sorghum sown after sunflower

g. Time of sowing of second crop also influence the choice of crops

If the second crop is sown in the standing first crop, the seeds will germinate before the harvest of first crop and may suffer some damage during the first crop harvest. Under such situation the crops which can with stand such damages in seeding stage must be selected. Eg. Rice fallow pulse.

h. Crops which may serve as alternate hosts

For certain pests and diseases alternate host crops must be avoided as succeeding crops.

Eg. Bhendi after cotton

i. Crops which make the best use of solar radiation

Crops which make the best use of solar radiation must be included to the extent possible. The solar energy use efficiency of intensive cropping systems will vary with the types of crops forming the crop sequence.

Eg. C_4 plants like maize, sugarcane, etc must be included, which are having rapid canopy development in summer when the solar radiation is high.

j. Short duration crops

Short duration crops like vegetables, cereals, pulses fodder crops are to be preferred to increase the intensity of cropping.

ii) Selection of varieties

For higher productivity - selection of varieties must be done with reference to their yield. Sensitiveness to photoperiod and duration - short and medium duration varieties which are photo insensitive to be selected.

a. *Duration: Rice fallow season:* In pulses, blackgram T9 or CO5, Cotton MCU 9 or SVPR 1 (135 days duration) can be selected. If the rice fallow crops are longer duration, it will interfere in the operations of first season rice.

b. *Photo sensitiveness:* GEB 24 Photosensitive variety and season round one comes to flowering only during Oct-Nov months. But now all are photo insensitive varieties.

c. *Yield:* Short and medium duration varieties which are photo insensitive and can be recommended for all seasons and yield higher productivity.

II. Selection of crops and varieties for intercropping

i) Selection of crops

Competition between component crops for solar radiation, water and nutrients. The possible changes in pests and diseases and weed incidence due to different spacing of crops that are grown together should have certain complementary effects of crop association. The factors that have to be considered for selection of crops and varieties for intercropping are as follows

a. Main crop yield does not suffer much

b. Competition between crop species is minimum

c. Complementary effects will be more

d. Pests, diseases or weed incidence is not unduly increased

e. Similarly the immobilization of N immediately after sorghum stalk less incorporated is another instance demanding proper residue management.

a. Competition for solar radiation in intercropping

In intercropping situations, competition for solar radiation might arise due to difference in height, leaf area, leaf orientation and growth duration.

i. *Height:* In intercropping systems, taller components are usually able to intercept more light and shorter components are affected by shading and low light intensities. If the intercrops are shorter they do not pose any competition for light with the main crop.

   Eg. In studies on sorghum based intercropping systems short growing crops like cowpea, soybean and greengram did not compete with sorghum for light and light interception by intercropped sorghum was more or less equal to sole sorghum. But when tall growing sunflower was intercropped in sorghum rows, light interception by sorghum was reduced by 50%.

ii. *Leaf area:* In a combination of tall and short crops, the amount of solar radiation available to short crops is very much reduced by shading of tall crops. If the shorter components have greater leaf area, leaf area / leaf weight ratio and adaptation to low light intensities, than they will survive better in such intercropping systems and the intercepted light will be of greater use.

iii. *Leaf inclination:* It refers to leaf orientation based on ideotype of plants. Regarding leaf inclination of component crops for intercropping, an ideal situation would be to have taller components with more erect leaves and the shorter components with more horizontal leaves. So, sufficient light falls on shorter components favouring maximum utilization. Such mixture would intercept and use more light and the competition effect would be minimal.

iv. *Growth duration:* When component crops have different growth duration, the peak demand for light would occur at different times. In maize + green gram system, peak light demand for maize occurs at about 60 days after sowing, by the time the green gram is ready for harvest. Greengram is able to make use of light available in early stages of maize growth, when shade is less.

Similarly in sorghum + redgram system, redgram grows very slowly in the early stages. At the time of sorghum harvest, redgram reaches its peak light demand period and at that stage shading by sorghum is less/nil.

b. Competition for nutrients in intercropping

Competition for various nutrients between component crops depends on root distribution pattern, growth stages of peak demand period, types of nutrients required more by each component crop.

i. *Root distribution pattern:* When root system of component crops overlap in the same zone, competition for nutrients would occur and the crop with more aggressive growth would suppress the nutrient uptake by the less dominant crop and affecting its gr1owth. So, there must be variation in rooting pattern of component crops.

ii. *Peak demand period:* Similarly, if the peak nutrient demands periods of the component crops occur at the same time then the competition will be high for the available nutrients. Such competition for nutrients can be minimized by selecting crops of different root distribution pattern. For instance, when a shallow surface rooting millet is intercropped with a deep rooted redgram then the roots of these two crops forage in different soil layers, thus avoiding competition.

iii. *Crop duration :* Similarly, if the inter crop has either less or more duration than the main crop, then the peak nutrient requirement periods of these two crops would occur at different times and thus competition at any particular time is less. The problem of nutrient competition can further be minimized by selecting crops which require more quantities of different nutrients. Eg. Legumes require phosphorus and millets demand nitrogen. The duration of the crops might decide the nature of intercrops. Eg. Sorghum + redgram from South West monsoon to North East Monsoon periods. Whereas, sorghum + cowpea for South West monsoon or North East Monsoon periods under rainfed lands.

c. Complementary effects of intercropping:

i. Certain crops require less light intensity and high relative humidity. Altered micro climate is provided when such crops are grown in between tall growing components in an inter cropping system.

   Eg. Turmeric/Ginger/Black pepper in coconut gardens.

ii. Similarly when legumes are intercropped with non-legumes of a lo1nger duration, N-fixed by legumes would benefit the non - legumes.

iii. Presence of rhizosphere microflora and mycorrhiza associated with one of the crops may lead to mobilization and availability of nutrients which may benefit the associated crop also.

iv. Similarly provision of physical support by a tall crop to a climbing type of intercrop is another example of complementary effect.

Eg. Coconut + pepper, sorghum + lablab, maize + climbing beans

v. Tall components may provide protection against wind for the short component crop

d. Allelopathic effects

Some crops may be unsuitable as intercrops because they secrete toxins into the soil which will adversely affect the associated crops. Roots of cucumber, leaves of *Eucalyptus globules,* decomposing residues of sunflower are known to produce allelo chemicals, affecting the growth of other crops. Such crops must be avoided.

e. Pests, diseases and weeds infestation

There are instances where certain species of crops when grown together result in more in a dense of particular pest/ disease/weeds affecting the other crops. Such combinations must be avoided.

Ultimately the final aim in selection of crops for intercropping must be

a. To ensure normal yields of main crop and additional yield from intercrop (or)

b. To produce a higher total yield from both crops than what either of them would have produced as pure crops

ii. Selection of varieties for intercropping

Even within a combination of two particular crops, selection of varieties of each crop suitable for inter cropping may become necessary. Genotypes suitable for sole cropping may not be suitable for intercropping situations and may be based on the following criteria.

a) differences in duration
b) differences in distribution of leaves on the stem
c) differences in rooting pattern
d) differences in growth habit

Eg. In sorghum + cowpea intercropping system, a sorghum variety of 100-120 days duration mixed with cowpea genotypes of 75-90 days duration and erect compact canopy without tendrils having deep root system than sorghum would produce better results.

III. Selection of crops for multi-tier cropping

Besides the criteria mentioned for intercropping other points to be considered are,

- Differences in canopy height to make use of solar radiation available at different heights from the ground level.
- Crops maturing in different periods so as to provide income at short and regular intervals in a perennial plantation
- Crops of different rooting pattern to avoid overlapping and to facilitate exploitation of larger and deep soil layers
- Crops which need shade and low temperature and increased humidity.

IV. Selection of crop for border/bund cropping

- Should not require extra care
- Should withstand damage due to passage of human and cattle
- Mature earlier or along with the main crop
- Should not create any border effects due to shade or root effect on the main crop

6.2 Tillage and Land Shaping

In sequential cropping, the main aspects to be considered are

a. When cropping intensity is increased i.e. more crops are raised in one year, the time interval between any two crops is very much reduced and this would affect the intensity of tillage and tillage operations are not carried out thoroughly.

b. Preparatory cultivation for second crop(s) is to be carried out and completed very quickly to enable sowing of the succeeding crops in time. Delayed planting of one crops would not only reduce its yield but also delay the sowing of the next crop and as a result the total productivity would also suffer.

In double crop rotation,

Eg. Sorghum - (March - June) : 110 days
Cotton - (August - January) : 165 days } 275 days

Here during summer after pulling out of cotton stalks, field can be ploughed leisurely for sowing sorghum crop and similarly there is one month gap after sorghum harvest for preparatory cultivation to the subsequent cotton crop.

For Triple crop rotation, i.e.

Sorghum (Feb - May) - 100 days
Ragi (May - August) - 90 days
Cotton (August - January) - 170 days } 360 days

Very little time is left for land preparation for ragi and cotton. If land preparation after sorghum is delayed, ragi seedlings would become aged and yield will be reduced. If ragi is planted late, it would delay the cotton sowing or affects its yield.

c. In some instances, to avoid delay in planting desired land shaping of the succeeding crop may not be possible before sowing and may have to be done late.

   Eg. Rice-fallow pulse/cotton

   Pulses have to be sown before moisture is last after the harvest of rice. Any normal preparatory cultivation like ploughing, forming beds would lead to soil drying and affect the germination and establishment of pulses. In rice fallow cotton thorough preparation of field for cotton including ploughing, forming ridges would invariably delay the sowing of cotton and result in delayed harvest of cotton affecting planting of first crop rice in the next season.

d. Increase in cost of cultivation due to increase in land shaping in the sequential cropping systems.

Management of tillage operations and land shaping can be done by the following means

a. To overcome the shortage of time between crops, use of implements and machinery which will reduce the time required for field preparation and land shaping can be resorted. Eg. Use of tractor drawn plough harrows, cultivators, cage wheel etc.

b. Animal / tractor drawn bed former, furrow opener, seed drills etc.

c. Two to three operations at a time (once over tillage) can be covered with certain implements. Animal drawn or tractor drawn seed drill – opens the soil, places the seeds, places the fertilizer and covers the seeds and fertilizers.

d. To check the increase in cost of operations use of labour saving implements may be followed. Eg. Bund former, Ridge plough, seed drill.

e. Minimal tillage to reduce the number of ploughings to the maximum possible.

f. Zero tillage – sowing crops without any preparatory cultivation

g. Chemical tillage and mulching

h. Off-season tillage

i. Land shaping after crop establishment

j. Relay cropping

k. Semi-permanent land shaping

l. Measures to reduce main field duration

a. Minimum tillage

It refers to restricting the number of tillage operations to the minimum possible required level to facilitate germination and establishment and also to restrict the tillage operations to the seeding zone only within the field.

For eg. Instead of 3-4 ploughings which is normally given, if reasonable tilth can be obtained within 1-2 ploughings, to that extent preparation can be completed early.

In most of the crops, the seeding zone has been defined as the area around the growing young plant about 20 cm deep and 10 cm diameter for widely spaced crops. For closely planted crops the seed zone would be 10 cm wide along the crop row. In the minimal tillage, only the seed zone is tilled intensively and the inter space is not ploughed or ploughed once. This would reduce the time required for tillage.

Many advantages are attributed to the minimal tillage

a) Cost reduction, b) Time saving, c) Reduced run off and erosion, d) Reduced evaporation, e) Reduced soil temperature.

However, in heavy poor drained soil, minimal tillage may not be effective.

b. Zero tillage

Soil is not disturbed mechanically and not opened up. For placing the seeds alone the soil is opened. Succeeding crop is sown, without any preparatory cultivation in the stubbles of the previous crop. Advantages are i) cost reduction, ii) time saving, iii) Reduced run off and erosion, iv) reduced evaporation, v) reduced soil temperature, vi) soil structure is improved.

For eg. Cotton is sown in ragi stubbles or paddy stubbles. Presence of fibrous roots and stubbles of ragi is believed to reduce surface encrustation and enable easy emergence of cotton seedlings, especially in heavy soils. Fallow cotton in paddy avoids delay due to field preparation and cotton growth can be completed before the next paddy season commences.

However, the zero tillage is not popular in India as in western countries where the labour is costly and not available, climate is temperate and there when organic matter is placed in the soil, decay is very slow and not rapid. There will be severe weed menace which can be controlled by wide range of herbicides. Under Coimbatore condition with three crop rotations, ragi-cotton-sorghum under zero tillage with chemical weed control, *Cynodon dactylon* became a major problem after the second year and was difficult to control. However, the ragi stubbles are soft with less cellulose fibre when subjected to decomposition and mineralization, encrustation of soil is prevented thereby improving the emergence of cotton seedlings.

c. Chemical tillage and mulching

In both minimal and zero tillage, the main problem is weed infestation and regrowth of crop stubbles, in the absence of thorough preparatory cultivation. This can be tackled by the use of suitable chemicals to control the weed growth and regrowth of stubbles of previous crop, which is termed as chemical tillage.

Eg. Application of paraquat in rice stubbles would reduce the seedlings regrowth from the stubbles and weed growth and dried residues will act as mulch. Similarly, pre-emergence herbicide for zero tillage plots. Mulching with crop residues in minimal/zero tillage plots helps to control weed growth.

d. Off season tillage

Tillage during non-cropping season in heavy soils, deep and intensive cultivation with tractor drawn disc plough or mould board plough during off season/dry season (summer ploughing) would help to break the hard pan and control perennial weeds like *Cynodon dactylon, Cyperus rotundus* and control pests like red hairy caterpillar of groundnut (pupal exposure). Minimal tillage during the cropping season would be enough and timely seed bed preparation is possible during cropping period favouring timely sowing of crops.

e. Land shaping after crop establishment

This is to overcome narrow time interval and not to miss the correct season of sowing. In some crops sown with minimal or no tillage, suitable land shaping can be done after the crop establishment.

Eg. Cotton sown in ragi stubbles, ridges and earthing up operations are carried out at the time of first top dressing of nitrogen at 30-35 DAS.

Similarly, forming ridges and furrows for rainfed cotton at 30-40 days after sowing to avoid delay in sowing season.

f. Relay cropping

There is a very quick succession of crops to avoid delay between crops, sowing the succeeding crops in the standing previous crop may be practiced without any land preparation. It saves time and reduces the cost for land preparation.

Eg. Rice fallow blackgram, 10-15 days before the harvest of paddy, pulse is sown.

Relay sowing of horse gram in standing crop of ragi under rainfed conditions of Salem and Dharmapuri and Pudukottai districts of Tamil Nadu.

Salkepai : dropping the ragi seedlings behind the country plough in furrows during South West Monsoon periods and horsegram sowing in North East Monsoon period, under raised conditions in North Western Zone of Tamil Nadu.

g. Semi permanent land shaping

Land is shaped not temporary but for a period of few years. In drylands, where double cropping is practiced i) to ensure timely sowing of second crop and ii) to overcome the problem of heavy rainfall, making it difficult for land preparation during the interval between two crops, seed beds may be formed on a permanent basis.

Eg. Broad Bed Furrow system (BBF) is a classic example of this practice. These are formed during off season and can be maintained for 3 to 4 seasons with light harrowing of seed beds only at intervals to kill the weeds, to close the kracks on the surface for sowing the succeeding crops. BBF system is a semi permanent structure. 120 cm bed width, 30 cm deep furrow, suitable for vertisols or black cotton soils, where double cropping is possible under rainfed conditions.

The land is thrown in to beds and furrows with a mild gradient of slope (0.4 to 0.8%) with the receipt of heavy rains, there will be water stagnation in black clay soils as it has heavy clay content (> 52 per cent). So, the soil can not be brought for early sowing of both first and second season crops, which may be delayed leading to crop failure. So, if the land is brought into beds and furrows with gentle gradient, the excess water is drained through many small channels, thereby preventing soil erosion. In addition soil will be friable which is required for early sowing.

In our state it is having limited applicability because of lesser black soil areas with heavy rains. But, highly suitable for MP and Maharashtra where sorghum -safflower/ sunflower/Bengalgram /mustard is being practiced.

Measures to reduce main field duration

- Short duration varieties
- Raising nursery and transplanting
- Hastening the maturity of crops
- Paraquat spray in cotton –boll bursting
- NaCl in rice-maturity is rapid
- Ripeners (Mont 8000) in sugarcane
- Harvesting at physiological maturity

  i.e. Seeds or grains will be having maximum vigour and viability but has more moisture than when the seeds are collected during normal harvest time. Similarly, in millets if harvested earlier, the fodder will be greenish.
- Relay sowing will reduce the main field duration of both crops in the sequence
- Ratooning
- Avoiding delay in maturity due to excessive vegetative growth. Eg.

Crops	Plant/sown crop duration (days)	Ratoon crop duration (days)
Sorghum	105-110	75-80
Paddy	110-115	80-95
Sugarcane	360	300

Topping in cotton and growth regulator application - CCC at 40 ppm

Topping and desuckering in tobacco

Reducing N supply in cereals and increasing the N supply in sugarcane will avoid delay in maturity.

With holding irrigation also will avoid delay in maturity.

6.3 Plant Population, Crop Geometry, Method and Time of Planting

The yield advantage in an intensive cropping system depends to a larger extent on the establishment of a good crop stand in which the component crops enjoy a favourable environment to perform well and increase the benefits of crop associations. Besides routine aspects like good seeds, correct and ideal seed bed preparation, optimum moisture in soil and removal of weeds, the establishment of a good crop stand also depends on optimum population of component crops, proper crop geometry, suitable method and time of sowing. The significance of these management aspects are relatively more important in intercropping than in sequential cropping since in the intercropping competition occurs both in space and time dimensions, affecting stand establishment. Therefore, there should be optimum population of both the component crops.

a. The planting pattern should be modified to avoid competition

b. The time and method of sowing also should be altered to minimize competition.

Distribution of plants over the ground area is called as crop geometry. It also refers to the shape of the land area available to the individual plants. The shape is altered by changing the row arrangement.

i. Optimum population of component crops

In the intercropping, when two or more crops are raised together, the total population of plants per unit area is normally more than that of the sole crop or base crop. The proportional population of each component must be carefully fixed in order to avoid competition leading to poor growth of either or both the crops and total yield advantage of intercropping system would be reduced.

Depending upon the objective and competition the population of the component crops will vary.

a. When a near full yield of the base crop is defined, the main crop population must be equal to that when it is raised as sole crop and intercrop population has to be lesser than under its pure crop. It may vary 30-50% of pure crop population. This is true especially when intercropping is also equally competitive.

b. In intercropping, if one of the crops is more dominant and competitive, the other crop is likely to suffer and its individual plant performance would be lower than when it is raised as pure crops. In order to maintain yield advantages and to make up the loss in individual plant performance, the intercrop population may be increased even upto 100 % of sole crop, provided the main crop is not affected.

c. In other situations, where a particular combination of crops is likely to result in only a total yield advantage or increase, but the performances of either of the crops is likely to be less than under sole crops, the proportionate population of component crops must be fixed so as to get the highest yield advantage.

ii. Crop geometry in intercropping system

After fixing the plant population, if sown as such then there will be a competition. Even in wide spaced crops, when intercrops are raised, suitable alternations in row arrangement should be made.

Sowing crops in the normally recommended uniform row spacing of pure crops would afford little or no opportunity for accommodating a companion cropping. On the other hand a modification of the planting pattern of the base crop would make intercropping feasible and often remunerative. This is made possible by keeping the plant population density per unit area constant, altering the orientation of rows with in certain limits, which does not result in any deviation in yield of the crops. Possible changes in crop geometry for the base crop in an intercropping system from that of pure crop would be

i. wider inter row spacing and reduced intra row spacing

   eg. 60 x 30 cm - Sole redgram

   90 x 20 cm - Intercropping system

ii. Paired rows of main crop or base crop

   eg. Sorghum 45 x 15 cm (or) 60/30 x 15 cm

iii. Uniform row planting with a replacement of main crop rows by intercrop rows

   eg. Sorghum + Black gram at 2:1 ratio

Examples

i. In sorghum, 45 cm inter row spacing is too close for raising an intercrop. If the row spacing is increased to 60 cm and plant spacing is reduced to 10 cm from 15 cm, the sorghum population would not be reduced much, and the inter space of 60 cm is sufficient for raising one row of inter crop.

Paired row planting of main crop provides adequate inter space for sowing intercrops

$$45 \times 15 \text{ cm} \rightarrow 60/30 \times 15 \text{ cm}$$

$$60 \times 30 \text{ cm} \rightarrow 90/30 \times 30 \text{ cm}$$

iii. In many intercropping system to accommodate intercrop one or more rows of main crops are replaced by intercrops

eg. Groundnut + Redgram - 6:1

Sorghum + Redgram - 8:1

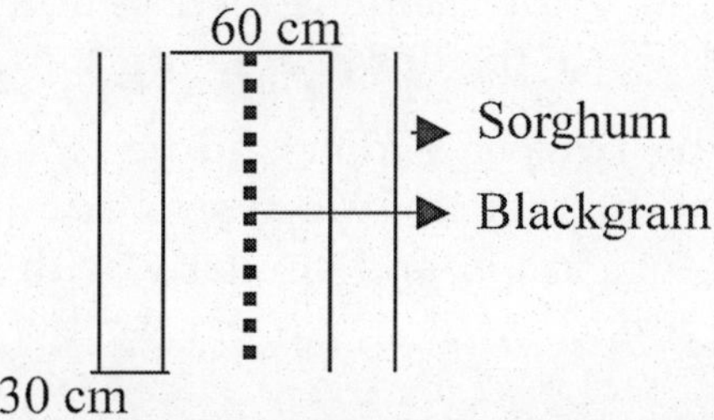

iii. Method of sowing in intercropping system

When broadcasting is followed, it will be difficult to ensure optimum population and proper crop geometry, which is necessary for success in intercropping. So drilling/ sowing behind the plough, dibbling are suitable.

To facilitate staggered sowing of intercrops one of the crops may be transplanted instead of direct seeding. Thus method of sowing should be chosen in such a way as to ensure optimum population, correct crop geometry and proper time of sowing of component crops.

iv. Time of sowing of intercropping system

Adjusting time of planting /sowing is an important criteria for avoiding competition between crops in intercropping, relay cropping and also in some sequential cropping system.

Normally in intercropping systems, both the main and intercrop are sown simultaneously for ease and convenience in sowing. This may not cause much problems if the crops differ widely in their growth pattern.

Eg. Sorghum + redgram, cotton + blackgram, groundnut + castor/redgram

If it is found that one of the crops is dominant and aggressive and may suppress the growth of the other, staggering in sowing time may be helpful. The less aggressive crop may be sown earlier and the dominant crop sown later. This would facilitate the less dominant crop to establish well before the dominant crop starts competing.

Eg. Sowing cowpea earlier and transplanting cumbu after 20 days, ensuring 20 days of competition free environment for cumbu and similarly with groundnut + sunflower, groundnut is sown earlier by a week than sunflower.

In some intercropping situations, the necessity for doing a particular cultural operation for the main crops in time may cause the sowing of the component crops at different times.

In cotton + onion intercropping systems onion is planted on one side of the ridges, 2 weeks before cotton sowing. This will enable onion to come to harvest at 45-50 days after sowing of cotton, at the earthing up stage of cotton. If two crops are simultaneously planted, onion will be ready for harvest only at 70 days and earthing up could not be done on 45-50 days after sowing.

Similarly, in most part of Karnataka, green gram / cluster bean is sown 13 weeks before cotton sowing.

In relay cropping situations, the period of overlap in growth persists between the two crops must not be too long to affect either of the crops. The relay crop must be sown before the harvest of previous crop when moisture is optimum for its germination at the same time, it must be sown in sufficiently later growth stages of first crop to avoid any competition effect.

Eg. Rice fallow pulse is sown 7-10 days before the harvest of paddy.

Horsegram is sown 30-35 days before the harvest of ragi.

In sequential cropping systems, if the previous crop stubbles are likely to result in allelopathic effect or wide C/N ratio effect sufficient time interval must be allowed for proper decomposition of such stubbles before the second crop is sown.

Eg. After sorghum, sunflower etc.

6.4 Water Management

Water management has to be studied with reference to

i. Total water requirement
ii. Scheduling of irrigation

i. Total water requirement

Knowledge of anticipated total water requirement of different types of intensive cropping systems would help us to select or choose the one that would suit to water availability situation.

In an irrigated condition, total water requirement would include the water lost through ET by various component crops, water needed for land preparation, water lost during conveyance.

In rainfed lands, total water requirement is the water lost through ET by component crops. Let us examine the total water requirement in some intensive cropping on water management.

a. Intercropping
i. Rainfed

Pure pearlmillet	159mm
Pure groundnut	196mm
Pearl millet + groundnut	228mm
Increase due to intercropping	32-69mm

ii. Irrigated

Irrigated	SW monsoon	Summer
Groundnut pure	562mm	698mm
Groundnut + Greengram	612mm	745mm
Increase due to intercropping	50mm	47 mm
Sorghum	419mm	424mm
Sorghum + cowpea	490mm	523mm
Increase due to intercropping	74mm	99mm

It is evident that introduction of a component crop in between the rows of main crop, naturally increases the total population and consequently increases the total water requirement. But this increase is very minimal. Comparing the minimal increase in total water requirement and the high total yield, the water use efficiency of intercropping is more i.e. the production per hectare cm of water is high.

b. Sequential cropping

When the number of crops raised in a field in one year increases the total water requirement of the cropping system also increases.

Eg. Wetlands	Rice-Rice-Rice = 3500 mm
	Rice-Rice-pulses = 2300 mm
	Rice-Rice-Ragi = 2750 mm
Garden lands	Cotton - Sunflower - Ragi = 1925 mm
	Cotton - Sorghum - Ragi = 1975 mm
	Cotton - Maize - Ragi = 1985 mm

Such information in total water requirement of different cropping system would be useful for choosing the suitable crop sequence based on water availability.

ii. Scheduling irrigation

Scheduling irrigation refers to method of irrigation and frequency of irrigation. It depends on total quantity of water required by crops.

a. Intercropping

In intercropping system, the total water requirement varies and critical stages of water requirement of component crops may not coincide. If one crop is irrigated based on its requirement, then the other crop may suffer due to excess or stress.

Eg. Cotton + blackgram, cotton needs dry spell in the first 20-25 days but this may affect blackgram. More number of irrigations may be necessary in intercropping than in pure crop, because of increased total water requirement. When the intercrop is sensitive to excess water it may be raised in bed furrow system.

Eg. Sugarcane + soybean

6.5 Weed Management

Intensive cropping systems necessitate changes in weed management practices because

a. Decrease/increase in weed incidence
b. Change in weed flora
c. Selection of suitable method of weed control
d. Restriction in inter cultivation operations

i. Intercropping

a. Increase / decrease in weed incidence

Complete coverage of interspaces by the foliage of an intercrop in between the main crop rows may suppress the growth of the weeds. The weed suppressing ability of intercrop depends on the type of crop variety, its rate of growth, its canopy development and planting pattern.

Raising a quick growing wide canopied intercrop can provide effective weed control in widely spaced crops like banana.

Eg. Banana + cowpea

Cropping	**Dry matter of weeds at 40 DAS (g/m^2)**
Redgram + sorghum	92
Redgram + pearlmillet	97
Redgram + cowpea	87
Redgram + field bean	96
	Dry matter of weeds at 60 DAS (kg/ha)
Sorghum pure	332
Sorghum + blackgram	316
Sorghum + cowpea	292
Sorghum + lablab	297

When banana + cowpea intercropping is followed the weed competition is reduced upto 70 DAS. However, increased weed infestation in intercropping is also reported. eg. Weed population was not affected between maize + groundnut and pure maize. Generally, fast growing habit, early canopy coverage and increased population of intercrops control weeds. Reduced weed infestation, reduces the labour requirement for weeding as well as reduces the number of weeding required.

b. Changes in weed flora

Changes in the weed species are noticed between pure and intercropping.

Eg. Sugarcane + cowpea / cucumber - more of sedges and grasses. Suppressed broadleaved weed growth.

Changes in weed species would necessitate changes in method of control (herbicide).

c. Method of weed control

i) Method of weed control through intercultivation with implements may be different and only manual weeding may be possible. eg. Cotton – intercultivator

ii) Herbicide selective to main crop may adversely affect the intercrop, thus a careful choice of herbicide is essential, that does not harm the component crops.

eg. Sorghum pure - Atrazine

Sorghum + pulse - Alachlor or Isoproturon

ii. Sequential cropping

The aspects to be considered are

a. Effect of continuous crop cover on weed species incidence

b. Presence of certain crops in sequence on specific weeds

c. Residual effect of herbicides on succeeding crop

In sequential cropping, where crops are grown continuously in a piece of land, annual weeds is suppressed and perennial weeds start dominating. This makes weed control more difficult.

Dominance of *Cyperus, Cynodon* has been noticed in intensive cropping system of three or more crops in a year. The inadequate tillage, often accompanies intensive cropping may also increase the weed incidence.

Residual effect of herbicides applied to previous crop may interfere with the germination and establishment of succeeding crop.

Eg) Atrazine @ 0.5 kg a.i./ha applied to sorghum affected establishment of succeeding legumes like greengram, gingelly, blackgram and cowpea.

Weed management in sequential cropping requires a proper combination of cultural and chemical methods, proper choice of herbicide and its quality are necessary. An integrated weed management based on cropping system approach has to be developed for different sequential cropping system.

i. Off season tillage with proper intercultural methods
ii. Necessary weed management techniques
iii. Usage of herbicide with little or no residual effect
iv. Manual weeding after the pre-emergence herbicide

6.6 Soil Fertility Management and Fertilizer Use

Even for individual crop itself, fertilizer scheduling is very difficult.

Eg. Paddy 100:50:50 kg NPK/ha

Sorghum 90:45:45 kg NPK/ha

Sugarcane 225:62.5:112 kg NPK/ha

To design fertilizer schedule for a productive crop, certain factors have to be considered like soil fixation, losses like leaching, and type of fertilizers used.

Designing fertilizer schedule for intensive cropping is still more complex because

i. The residual fertilizer effect may be accounted. It may benefit certain crops in the cropping system.

ii. Legume effect is to be accounted. Due to decaying nodules, some amount of nutrients is released, accordingly, the fertilizer must be rescheduled.

iii. Competitive and exhaustive effect also to be accounted. Eg. Cotton removes more of phosphorus.

iv. Contribution from the stubbles and residues of previous crops/ component crops is to be considered.

When we design a fertilizer schedule based on cropping system, there is a possibility of reducing the fertilizer doses. The efficient fertilizer management is possible by

i. Nutrient requirement of various component crops should be known. The sum of the nutrient requirement of various crops is the total nutrient requirement of the system.

ii. Nutrient uptake pattern based on the nutrient demand of each crop should be known

iii. Response of crops to various nutrient application. Some crops may respond to a particular nutrient, but not the other crops for the same nutrient

    Eg. Pulse respond to phosphorus

    Millets, cereals respond to nitrogen

    Nuts tubers and oilseeds to po1tassium

iv. Rates and method as well as time of fertilizer application.

Mobile nutrients can be extracted more or less quantity from the soil volume surrounding the roots (called root system sorption zone) but relatively immobile forms can be extracted only partly depending on their mobility from the soil volume immediately around the roots (called root surface sorption zone). The fertilizer supply for mobile nutrients must be equal to the demand by crop whereas, for immobile nutrients based on the relative sufficiency of soil supply.

v. Integrated nutrient management for the cropping system by accounting the soil contribution and involving organic forms like FYM. Compost and residues contribution, inorganic forms and biofertilizer like Azospirillum for cereals, millets, cotton, sugarcane and Rhizobium for legumes and phosphobacteria for mobilizing the phosphorous.

6.7 Plant Protection Measures

Intensive cropping systems like intercropping and sequential cropping may increase / decrease the incidence of certain pest and diseases.

i. Increase in vegetative cover

ii. Alteration in micro-climate

Eg. Sorghum + pulses reduce the early shoot borer incidence

Sterility mosaic increases in Redgram + sorghum than in pure redgram due to micro climate changes

iii. Leaf miner in groundnut reduced when intercropped with cowpea. Castor as trap crop in cotton field.

iv. Presence of physical barrier affecting the movement of insects. eg.) Sorghum + sesame, less damage of web worm to sesame

The pest and disease management strategy include

a) Choice of resistant varieties
b) Careful combination of crops in intercrop
c) Careful combination of crops in sequential crop
d) Destruction of crop residues
e) Cultural operation for pest and disease control
f) Optimum time of sowing
g) Judicious use of selective pesticide

6.8 Labour Management

In crop production, land, labour and capital are three important resources to be managed efficiently. However, management of labour is a difficult process. Whenever, any programme is drawn the role of labour should also be considered. The present day share of labour in most of cost of cultivation of crops is 50-60%.

Energy requirement for various operations and the human labour energy utilization is also considerably high. Cost of human energy when compared to total energy is very much considered.

Crop	% of human energy to total energy
Cotton	2.15%
Bajra	17.65%
Maize	16.29%
Ragi	27.76%
Sorghum	13.10%

Efficient labour management is essential, but yet it is very difficult because labour as a resource has certain unique features which are not seen in other resources. They are

a. Services of labour cannot be stored. As and when available it is to be used.

b. Labour comes in indivisible unit

c. In small farms family labour constitutes a considerable share

d. Management of human labour requires specific skill involving human psychology.

Problems in management of human labour

i. Labour inefficiency

ii. Frequent increases in wage rates - not hoped by the prices of the produces

iii. Labour strike as a result of labour organization.

iv. Demand for wage hike during critical work periods

v. Shortage of labour during peak requirement period. i.e. there will be a shortage of labour in one part of the year, as agricultural is highly seasonal.

vi. Heterogeneous group - leads poor efficiency and quality of work is also affected.

Measurement of labour efficiency

1. Labour cost / unit area
2. Number of labours / unit area
3. Labour output / unit time
4. Productivity of crop / man day or man hour

i. Labour cost / unit area

It varies from crop to crop. Normal cost for a crop is Rs.1000/-, but the labour cost is Rs.850/- then there is labour efficiency. If it is Rs.1000 then there is labour inefficiency.

ii. No. of labours / unit area

For a crop, 200 man days normal. If it is > 200-poor labour efficiency, if it is <200, then better labour efficiency. It also varies from crop to crop.

iii. Labour output / unit time

It refers to work turned out by a labour in unit time or a day of 8 hours. It is assessed for each and every operation.

iv. Mandays or man hour

No. of man days / man hour required for a quantity of produce obtained / unit area. If the value is higher, then is desirable. There are certain limitations when this is used to assess the labour efficiency. With the same amount of labour, higher productivity could be obtained with efficient management of other inputs.

Methods of efficient labour management

The aim is

i. To increase the labour productivity

ii. To reduce the labour requirement

iii. To reduce the labour cost per unit area

iv. To overcome, the difficulties arising out of shortage of labour which in turn is caused during peak period of labour demand

v. Better personnel management

Methods to increase labour efficiency

i. To increase the labour productivity

a. By proper selection of labourers - with good physique, necessary skill and enough experience in various farm operations

b. By prompt payment of optimum wages

c. By carrying out the operations under optimum field conditions to increase the labour productivity.

eg. Weeding optimum stage and moisture

d. Right choice of tools
   Weeding with hand hoe or dryland weeder
   Maharashtra sickle for paddy harvest

ii. To reduce the labour requirement

a. Cost of production can be reduced.

During shortage of labour, it is a welcoming relief to the farmers as using labour saving implements

Eg. Forming ridges and furrows by ridger, it reduces 70-80% of labour requirement. By using 5 tynes seed drill for sowing bold seeds, 45% of labour can be saved, for green fodder cutting - power operated chaff cutters - 2 men labours can chop 5 tonnes of green fodder and 2 tonnes of dry fodder.

b. Using herbicides for weed control

For most crops, use of herbicides, reduce the labour requirement

Eg. For hybrid maize - Pre-emergence use of atrazine + late inter cultivation reduced the labour requirement.

c. By mechanization eg. Tractor for preparatory cultivation

Paddy thrasher - very quick with 5 man days coverage whereas in case of manual labourers, 26 man days are required.

d. By changing the method of crop establishment

Eg. By direct seeding, instead of nursery and transplanting, labour requirement can be reduced.

iii. Reducing the labour demand during peak period

Agricultural operations are seasonal and during the peak labour requirement period, labour shortage is felt due to labour intensive operations are overlapping.

Ways and means the labour peaks are

1. Allow willing labourers to do extra time work and pay extra wages.
2. Use of uncertain seasonal labour force. eg. Engaging school students for flower picking, stripping / shelling groundnut pods.
3. Adopt contract system of work. Payments are made on quantity of work turned out, again the problem is quality of work.
4. By changing the cropping pattern the sequence of crop should allow long distribution of labourers, without peaks and snags. There are certain crops for which season is not so rigid Eg. sunflower - day neutral plant.
5. Drawing of calendar of labour requirement and labour availability
6. Carrying out the less important operations during slack period Eg. Paddy harvest, thrashing, winnowing and cleaning, staking paddy straw and winnowing of dried paddy.

Tips for efficient labour management

1. Motivate the labourers by appreciation
2. Clear communication with crystal clear instructions
3. Try to fulfill their physical needs like shelter, food and clothing
4. Provide safety and security with better care and affection
5. Protect the self esteem of the labourers and also the achievement

The efficiency of crop production itself to a larger extent depends on the efficiency of labour management *"Listen to labourer"*

7. COST REDUCTION IN CROP PRODUCTION

Improvement in management practices help to realize higher yields but at the same time most of the production technologies are also input intensive and labour intensive. Costs of labour and input are higher and are increasing in upward trend at very frequent intervals. Increase in wages and inputs act as dhampers (Make farmers to loose their interest) from going for achieving higher yields. Therefore the motivation of the farmers to aim for higher productivity may be reduced if the profits are going to remain the same or sometimes less because of higher cost of cultivation. So, it is necessary to sustain higher yield levels at reasonable cost of cultivation.

In most of the crops, out of total cost of production, about 50-60% goes for labour wages and about 40 to 50% goes for input costs.

For example in rice cultivation, out of 100% total cost of cultivation.

Human labour	-	40%
Bullock labour	-	7%
Seeds	-	8%
Fertilizers	-	26%
Pesticides	-	5%
Irrigation	-	12%

Any attempt to reduce the cost should not result in reduced yields

Cost reduction in crop production can be achieved through

a. Improving the labour efficiency and reducing the cost of labour for various operations.

   Eg. Ridge formation for cane (40 men labour) - Rs. 1600

   Ridge plough (2½ pairs) - Rs. 250

   Rectification (10 labour) - Rs. 400

b. Reducing the levels of inputs used without affecting the yield

   Eg. Neem cake coated urea for rice there by nitrogen use efficiency is increased with reduction in N requirement

c. Adoption of low – cost technology

   Eg. Use of biofertilizer like azospirillum for cereals, millets, cotton, sesame and rhizobium for pulses

d. Proper management of non-monetary inputs or no cost technology

Non monetary inputs are defined as those cultural operations which help to achieve high yield at no extra cost and whose cost does not change with the level of output.

7.1 Non Monetary Inputs in Crop Production

The following are the some of the non-monetary inputs in the crop production

i. Selection of suitable varieties according to the region, season, problem soils etc.

   a. For low temperature - MDU 2 rice
   b. Saline soil - CO 43 rice
   c. During Samba - IR 50 rice

ii. Use of quality seeds without admixtures and to avoid gap filling

iii. Timely land leveling and shaping for efficient water and nutrient management in garden and dry land regions

iv. Optimum time of sowing / planting. Eg. Sowing of cotton during August 15th Turmeric - End of May

v. Optimum plant population. Eg. Soybean 3.33 lakhs / ha

vi. Optimal depth of sowing / planting. Eg. Rice 4-5 cm, Ragi 2-3 cm, Sorghum 3-4 cm, cotton, maize, groundnut 5-7 cm.

vii. Correct age of seedlings for transplanting (aged seedlings should be avoided)

Eg. Ragi/cumbu/sorghum 16-18 days

Paddy short duration varieties – 21 days

Medium / Long duration varieties – 30 days

viii. Timeliness in important field operations. Eg. Weeding during the critical crop weed competition period Eg. Rice 10-40 days after transplanting.

Maize 10-35 Days after sowing, Sugarcane 21-90 days after planting, early weeding reduces the labour cost

ix. Irrigation at proper time and at optimum level to avoid stress at critical periods.

x. Time of fertilizer application to coincide with peak nutrient demand period. Eg. Sugarcane - with 90 Days after planting. Beyond 90 days reduces the sugar recovery by N application

xi. Timely harvest to avoid harvest losses

xii. Proper care in post harvest processing and storage to avoid wastages. Eg. Drying to optimum moisture is necessary in groundnut to avoid aflotoxin problems.

7.2 Low Cost Technology

Techniques which involve very little cost but help to achieve higher yields in crop production.

1. Seed / Seedling treatments for pest and disease control
2. Bio-fertilizers for various crops and to soils
3. Seed hardening and seed treatment for convenience in sowing. eg. Acid delinting in cotton

4. Fertilizer use in nursery and fertilizer application
5. Placement of fertilizer like N and also treated materials
6. Application of growth regulators / promoters

Measures for reducing labour and inputs

1. Tillage, land shaping, intercultivation

- Minimal tillage, zero tillage, off season tillage
- Use of labour saving implements / machinery.

  Eg. Ridger, landformer, Ridger reduces the labour requirement by 60%.
- Use of correct type of tools.

  Eg. 5 typed seed drill – cost of sowing reduced by 45%
- With groundnut stripper 300 kg/day can be stripped, whereas with women labourers, 60kg/day only can be stripped.

2. Seeds and sowing

- Seed treatment
- Seed drill
- Relay sowing and ratooning

3. Fertilizers management

- Soil test based fertilizer recommendation
- Cropping system approach
- Bio-fertilizers
- Correct time and method of application
- Form of fertilizer1s to suit the soil condition and to reduce the losses
- Balanced use of fertilizers
- Integrated nutrient management

4. *Plant protection*
   - Seed treatment (Pest and disease management)
   - Protection in nursery stage
   - Pesticide spray at ET levels of damage
   - Biological control based on crop rotation and cropping system
   - Summer ploughing
   - Integrated pest / disease management

5. *Weed management*
   - Early weeding
   - Use of correct tools which will increase the labour efficiency
   - Line sowing / planting for intercultivation
   - Pre-emergence herbicide with little residual effect

6. *Harvest and processing*
   - Timeliness
   - Use of machinery

8. EVALUATION INDICES OF CROPPING SYSTEM

It is grouped under 4 major categories

i. Indices based on land use efficiency
ii. Indices based on biological potential.
iii. Indices based on economic viability.
iv. Indices based on energetic approach

8.1 Indices Based on Land Use Efficiency

i. *Multiple Cropping Index (MCI):* By Dalrymple (1971)

It measures the sum of areas planted to different crops and harvested in a single year divided by total cultivated area times 100.

$$MCI = \frac{\sum_{i=1}^{n} ai}{A} \times 100$$

Where,

n = Total no. of crops

ai = area occupied by i^{th} crop

A = total land area

ii. *Cropping Intensity Index (CII):* By Menegay *et. al.* (1978)

It assesses a farmer's actual land use in area and time relationships for each crop or group of crops compared to the total available land area and time, including land temporarily available for production.

$$CII = \frac{\sum_{i=1}^{N_c} ai\ ti}{\sum_{i=1}^{M} A_j \quad T_j}$$

Where,

N_c = Total no. of crops grown during time period T

A_i = area occupied by i^{th} crop

ti = duration occupied by i^{th} crop (months that crop occupied ai).

T = time period (usually one year)

A_o = Total cultivated land area available for use during T

M = Total no. of fields temporarily available during T

A_j = land area of jth field

T_j = Time period when Aj is available

Efficient cropping zone is judged by CII and LER.

iii. Cropping intensity / Intensity of cropping (CI)

$$CI = \frac{\text{Total cropping area}}{\text{Net cultivated area}} \times 100$$

$$CI = \frac{\text{Area under } kharif + rabi + zaid \text{ crops}}{\text{Area under actual cultivation}} \times 100$$

iv. Rotational Intensity (RI)

$$RI = \frac{\text{No. of crops grown in a field}}{\text{Years of rotation}}$$

e.g., Maize - potato - onion : 1 year

$$RI = \frac{3}{1} \times 100 = 300\%$$

Similarly: Maize - sugarcane - onion : 2 years

$$RI = \frac{2 \times 100}{2} \times 100 = 100\%$$

v. *Specific Crop Intensity Index (SCII):* by Menegay *et. al.*, (1978)

It is derivative of CII and determines the amount of area-time denoted to each crop or group of crops compared to the total time available

$$SCII = \frac{\sum_{k=1}^{N_k} a_K t_K}{AoT \sum_{j=1}^{M} A_j T_j}$$

N_k = Total no. of crops during T

a_k = area occupied by k^{th} crop

t_k = duration of kth crop

Model:

Area : 1.5 ha. Time - 12 months

AoT = 1.5 x 12 = 18 ha month

Temporarily available : 1.5 ha

0.5 ha x 4 months = 2.0
0.3 ha x 5 months = 1.5
0.7 ha x 6 months = 4.2

7.7

Temporarily available
1.5 ha Permanent field

0.75 ha	=	Sugarcane		0.75 x 12 months	=	9.00
0.75 ha	=	Rice	=	0.75 x 3.5 months	=	2.625
		Cotton	=	0.75 x 5.0 months	=	3.750
		S.D. Veget.	=	0.75 x 3.0 months	=	2.250

17.625

Temporarily available

0.5	-	Rice	=	0.5 x 3.5 months	=	1.75
0.3	-	Vegetable	=	0.3 x 4.0 months	=	1.20
0.7	-	Cotton	=	0.7 x 5.0 months	=	3.50

6.45

$$\frac{24.075}{18 + 7.7} = \frac{24.075}{25.7} = 0.94$$

CII = 0.94

It is used for vegetable intensity index, rice intensity index, field crops intensity Index etc.

vi. Relative Cropping Intensity Index (RCII)

It is the modification of CII and determines the amount of area-time allotted to one crop or group of crops relative to the area-time actually used in the production of all the crops.

$$RCII = \frac{\sum_{k=1}^{N_k} a_K t_K}{\sum_{j=1}^{N_c} a_{ki} t_i}$$

It is used for classifying farmers *viz.*, when relative vegetable Intensity Index is more than 50% then farmer will be called vegetable grower.

vii. Cultivated Land Utilization Index (CLUI): By Chuang (1973)

It is calculated by summing the products of land area planted to each crop, multiplied by the actual duration of that crop and divided by the total cultivated land area times 365 days.

$$CLUI = \frac{\sum_{i=1}^{n} ai\ di}{A \times 365}$$

i = 1, 2, 3 crops 1...........n.

n = Total no. of crops

a = Area occupied by i^{th} crop.

d = Days that i^{th} crop occupied ai.

A = Total cultivated land area available during 365 days period.

Model: Farm Size: 2.5 ha

Crop	Area/ ha	Duration / days	a.d.
Rice	1.00	100	100
Rice	1.75	120	210
Cotton	1.75	140	245
Sugarcane	0.75	365	275
Total			830

$$\frac{830}{2.5 \times 365} \times 100 = 90\% \text{ or } 0.9$$

If less than 100% it indicates that some area is left fallow.

If CLUI >1, land is completely cultivated

If CLUI <1, land is left fallow

Limitation

- Useful for relay cropping.
- It takes only the land available with farmer.

viii. Simultaneous cropping index (SCI)

It is computed by multiplying the HDI with 10,000 and dividing the product by multiple cropping index (MCI).

$$SCI = \frac{HDI \times 10{,}000}{MCI}$$

ix. Area time equivalent ratio (ATER)

It takes into account the duration of crops and permits an evaluation of crops on yield per day basis. It is a modification of LER.

$$ATER = \frac{LA \times DA + LB \times DB}{T}$$

Where, LA and LB are relative yield or partial LER of component crops. A and B, DA and DB are duration of crops. A and B, and T are the total duration of the intercropping system.

x. Staple land equivalent ratio (SLER)

In situations, where the primary objective is to produce fixed yield of one component (staple) and some yield of other crop, the concept of SLER is proposed.

$$SLER = \frac{MDA}{SA} + P \left\{ \frac{MB}{SB} \right\}$$

Where, MDA is derived yield of A in mixture yield and P, the proportion of land devoted for intercropping. The two crop yields in the intercrop mixture are MA and MB and the yield of crops grown as sole crop are SA and SB.

8.2 Indices Based on Biological Potential

Crop Equivalent

One other form of single measurement comparison which is exactly equivalent to the financial value index, is the crop equivalent. In calculating a crop equivalent, yield of one crop is converted into yield equivalent of the other crop by using the ratio of prices of the two crops. Indices based on Biological potential are categorized into production efficiency index and Interference indices.

8.2.1. Production Efficiency Index

i. LER (Land Equivalent Ratio)

It denotes relative land areas under sole crop required to give the same yield as obtained under a mixed or inter cropping system at the same level of management. It is the ratio of land required by pure crop to produce the same yield as that of intercrop.

$$\text{LER} = \frac{Ya}{Sa} + \frac{Yb}{Sb}$$

Where,

Ya and Yb = Yield of individual crops a & b respectively in mixture.

Sa & Sb = Yields of individual crops a & b respectively in pure stand.

Thus, it gives a better picture of the competitive abilities of the component crops. It also gives actual yield advantage of intercropping.

In another words LER is the unit to measure the production efficiency of different intercropping system by convening the production in terms of land acreage. LER gives an accurate assessment of the biological efficiency of intercropping.

When LER > 1, intercropping is beneficial

ii. Relative Yield Total and Land Equivalent Ratio

The most important index of biological advantage is the relative yield total (RYT) introduced by De Wit and Van Den Bergh (1965) or land equivalent ratio (LER) reviewed by Willey (1979). The mixture yield of a component crop expressed as a portion of its yield as a sole crop from the same replacement series is the relative yield of the crop and the sum of the relative yields of component crops is called relative yield total (RYT). The total land area required under sole cropping to give the same yields obtained in the intercropping mixture is called land equivalent ratio (LER). Both the expressions (RYT and LER) are similar.

Example

Yield of groundnut and pigeon pea as sole crops were 1000 and 500 kg ha^{-1}. Corresponding yield in intercropping was 700 and 400 kg ha^{-1}. Price of pigeon pea is Rs.12 kg-I and that of groundnut Rs.10 kg^{-1}. Calculate LER.

Solution

$$\text{LER} = \frac{700}{1000} + \frac{400}{500}$$

$$= 1.5$$

Other indices have been proposed as measures of biological performance. There are two different objectives for which such indices have been proposed. The first is the assessment of the benefit or overall advantage of intercropping and the second is the assessment of the relative performance of the two crops, the concept of dominance or competitiveness. It is important not to confuse these two objectives, which should be quite separate conceptually. The RYT or LER is the main index of advantage currently used.

8.2.2. Interference Efficiency

i. Relative Crowding Coefficient (RCC): Proposed by De Wit (1960)

It is used in replacement series of intercropping. It indicates whether a species or crop, when grown in mixed population, has produced more or less yield than expected in pure stand.

In 50 : 50 mixture

$$\text{Kab (RCC)} = \frac{\text{Mixture yield of a}}{\text{Pure stand yield of a} - \text{Mixture yield of a}}$$

$$= \frac{Yab}{Yaa - Yab}$$

Yab = mixture yield of a crop grown with b

Yba = mixture yield of b crop grown with a

Yaa = yield in pure stand of crop a

Ybb = yield in pure stand of crop b

Zab = proportion of sown species a in mixture with b

Zba = proportion of sown species b in mixture with a

For all mixture

$$Kab = \frac{Yab \times Zba}{(Yaa - Yab)\ Za}$$

K > 1 Means yield advantage (more yield than expected)

K =1 No difference

K <1 Yield disadvantage (less yield than expected)

Crowding coefficient and LER give the yield advantage but only LER give the magnitude of advantage. Therefore LER is preferred to assess the competition effects and yield advantage in intercropping situations.

ii. Aggressivity

Proposed by Mc Gihrist (1965). It is the mixture of how much the relative yield increase in component a is greater than that for component b.

$$Aab = \frac{\text{Mixture yield of a} \quad \text{mixture yield of b}}{\text{Expected yield of a} \quad \text{Expected yield of b}}$$

$$Aab = \frac{Yab}{Yaa \times Zab} - \frac{Yba}{Ybb \times Zba}$$

Aab = 0 means component crops are equally competitive

Aab = negative means dominated

Aab = bigger value either positive or negative means bigger difference in competitive abilities.

Aggressivity and crowding coefficient can be used only in replacement series whereas LER in both replacement and additive series.

iii. Competition Index (CI)

Proposed by Donald (1963). It is a measure to find out the yield of various crops when grown together as well as separately. It indicates the yield per plant of different crops in mixture and their respective pure stand on an unit area basis.

If the yield of any crop, grown together is less than its respective yield in pure stand then it is harmful association but on increased yield means positive benefit.

$$CI = \frac{(Yaa - Yab) \times (Ybb - Yba)}{Yaa \times Yab}$$

iv. Competition Ratio (CR)

Proposed by Willey and Rao (1980)

$$Cra = \frac{Yab}{Yaa \times Zab} + \frac{Yba}{Ybb \times Zba}$$

$$= \left\{\frac{Yab}{Yaa} + \frac{Yba}{Ybb}\right\} \times \frac{Zba}{Zab} = \frac{LERa}{LERb} \times \frac{Zba}{Zab}$$

It is simply the ratio of individual LERs of the two component crops, but correcting for the proportion in which they were initially sown.

v. Competition Coefficient (CC)

Ratio of the Relative crowding coefficient (RCC) of any given species in the mixture.

$$CC = \frac{\text{RCC of given spp}}{\text{Total RCC of all crops in mixture}}$$

It is used to find out the relative crowding from which maximum yield can be obtained without any adverse effect on any of the species

8.3 Indices Based on Economic Viability

i. *Gross return:* the total monetary returns of the economic produce such as grain, tuber, bulb, fruit, etc and by products viz., straw, fodder, fuel, etc obtained from the crops included in the system are calculated based on the local market prices. The total return is expresses in terms of unit area, usually one hectare.

ii. *Net return or net profit:* Worked out by subtracting the total cost of cultivation from the gross returns. This value gives the actual profit obtained by the farmer. In this type of calculation only the variable costs are considered

iii. *Returns / rupee invested:* Also called as benefit cost ratio or input output ratio.

$$\text{Return per rupee invested} = \frac{\text{Gross returns}}{\text{Cost of cultivation}}$$

iv. *Per day return:* Also called as income per day and can be obtained by dividing the net returns by number of cropping period (days)

$$\text{Per day return} = \frac{\text{Net returns}}{\text{Cropping period (days)}}$$

v. *Diversity Index (DI):* By Strout (1975) and Wang and Yu.

It measures the multiplicity of crops or farm products which are planted in a year by computing the reciprocal of sum of squares of the share of gross revenue received from each individual farm enterprise in a single year.

In the following farms, which farm is mostly specialized.

Crops	Income (Rs.) From Different Farms		
	A	B	C
Sugarcane	30,000	-	10,000
Cotton	10,000	20,000	20,000
Wheat	40,000	29,000	10,000
Jowar	20,000	10,000	40,000
Potato	-	50,000	-
Total	100,000	100,000	80,000

Solution

Crops	Shares of individual crops in different Farmer Participatory research on Integrated Farming System in Low External Input Sustainable Environment					
	Farm A		Farm B		Farm C	
	Share	Square of its share	Share	Square of its share	Share	Square of its share
Sugarcane	0.3	0.09	-	-	0.125	0.0156
Cotton	0.1	0.01	0.2	0.04	0.25	0.625
Wheat	0.4	0.16	0.2	0.04	0.125	0.0156
Jowar	0.2	0.04	0.1	0.01	0.50	0.2500
Potato	-	-	0.5	0.25	-	-
Total	1.0	0.30	1.00	0.34	1.00	0.3437

Diversity Index

$$\text{Farm A} = \frac{1.00}{0.30} = 3.33$$

$$\text{Farm B} = \frac{1.00}{0.34} = 2.94$$

$$\text{Farm C} = \frac{1.00}{0.3437} = 2.91$$

Therefore, Farm 'A,' is more diversified and Farm C is more specialized

Lower is the diversity Index, higher is the specialization.

vi. Harvest Diversity Index (HDI)

It is computed using the same equation as the DI except that the value of each farm enterprise is replaced by the value of each harvest.

$$HDI = \frac{1}{\sum_{i=1}^{n} \left\{ y_i \sum_{i=1}^{n} y_i \right\}}$$

where, y_i= gross value of the ith crop planted and harvested within a year.

vii. Monetary advantage based on LER

$$\text{Value of combined intercrop yield} \times \frac{LER}{LER} - 1$$

S. No	Crop	Yield	Price	Value
1.	Sole groundnut	1185	8.0	9480
2.	Intercrop groundnut	840	8.0	6720
3.	Sole millet	2276	2.0	4552
4	Intercrop millet	1227	2.0	2454

$$\text{Monetary advantage} = 9174 \times \frac{0.26}{1.26} = 1893.0$$

8.4 Indices Based on Energetic Approach

$$\text{i. Energy efficiency} = \frac{\text{Energy output (MJha}^{-1}\text{)}}{\text{Energy input (MJha}^{-1}\text{)}}$$

ii. Net energy ($MJha^{-1}$) =
Energy output ($MJha^{-1}$) - Energy input ($MJha^{-1}$)

$$\text{iii. Energy productivity (kg MJ}^{-1}\text{)} = \frac{\text{Output (grain + byproduct) (kg ha}^{-1}\text{)}}{\text{Energy input (MJha}^{-1}\text{)}}$$

$$\text{iv. Energy intensity (in physical terms, MJha}^{-1}\text{)} = \frac{\text{Energy output (MJha}^{-1}\text{)}}{\text{Output (grain + byproduct) (kg ha}^{-1}\text{)}}$$

$$\text{v. Energy intensity (in economic terms, MJ Rs.}^{-1}\text{)} = \frac{\text{Energy output (MJha}^{-1}\text{)}}{\text{Cost of cultivation (Rs. ha}^{-1}\text{)}}$$

Questions

1. Explain the need for cropping system research
2. Classification of multiple cropping with examples

3. Explain plant interactions in cropping system
4. Write in detail about crop rotation and their advantages
5. Enlist and explain the agronomic requirements of intercropping system
6. Write a note on different types of tillage practices
7. Give the techniques for reducing labour and inputs
8. Explain the methods of labour management
9. Give the categories of cropping system evaluation and write about the indices based on biological potential
10. Explain about non monetary inputs and low cost technologies in rice based cropping system

—x—

Chapter 2

Feed and Forage Crops

1. LIVESTOCK AND INDIAN AGRICULTURE

Agriculture in this country is oriented towards mixed farming, where livestock rearing is an integral part of it. Livestock have become the bred winner of the millions of rural Indians. Indian agriculture is largely an economic symbiosis of crop and cattle production. It also offers the most efficient and environment friendly way for recycling of natural resources. On one hand, economic and efficient utilization of farm by-products and wastes by the animal and on the other hand, production of the most valuable manure for the farm lands to enhance productivity. Grazing based animal husbandry is important in rural India. The higher production potential of the crossbred animals and its economic sustainability has its main stay on judicious feeding round the year. The low productivity of Indian livestock is mainly due to chronic shortage of feed and fodder coupled with poor quality.

1.1 Forage Production Problems and Perspectives

Though India supports nearly 1/5 of the world's cattle population (with only 2% of world's geographical area) we are far away from, self sufficiency in milk, meat, wool and other animal products. Milk production is a factor of genetic potential, nutrition, husbandry, management and health. The critical analysis of the complex reasons reveals the fact that there is a big gap between the actual need and the provisions made to the cattle. Looking into the overall scenario, one can easily see that the animals are not provided with sufficient nutrition, as a result they are not able to show their genetic potentiality.

In the economics of milk production, nearly 62 – 65% of the total cost of milk production is due to the feeding of animal; when it is fed on fodder based ration the cost could be brought down to 40% but on the concentrates oriented feeding the cost structure raises to 80% of the production cost. Thus it needs no emphasis that feeding of animal is a single factor which can be identified as the major constraint in economic livestock production. Addition of green fodder of both non-legumes and legumes helps to reduce the cost of concentrate feeds as well as maintains the normal health and reproduction of all ruminants.

Table. 1. Livestock pressure - India

Livestock	Population (million)	Per cent
Cattle	198.4	44.36
Buffalo	77.0	17.22
Sheep	55.7	12.45
Goats	112.0	25.04
Horses and ponies		
Camels, Mules	4.1	0.93
Total	**447.2**	**100.00**

The fodder production is 1015, 600 and 90 million tones of green fodder, dry fodder and concentrates, respectively. The corresponding deficit is estimated to be 64, 16 and 80%. With the present feed and fodder resources we are able to meet only 46.6% of the animals requirement and that is why we are getting only 50% of desired animal products.

To attain the goals of rearing hybrid cows, feeding them with nutritious fodder is the most important aspect which improves the health of the animal. The heifers will conceive between 15 - 18 months and calve by 25 - 30 months. Again they may calve by 12 - 14 month intervals and give a good milk yield.

Quality of ration is determined by dry matter intake generally 1/3 of dry matter (feed) will consist of concentrates, 2/3 rd of dry matter will be through roughages - green or silage, hay or other dry roughages. *Eg.* A cow weighing 400 kg consumes 3% of body weight requires about 12 kg of dry matter per day.

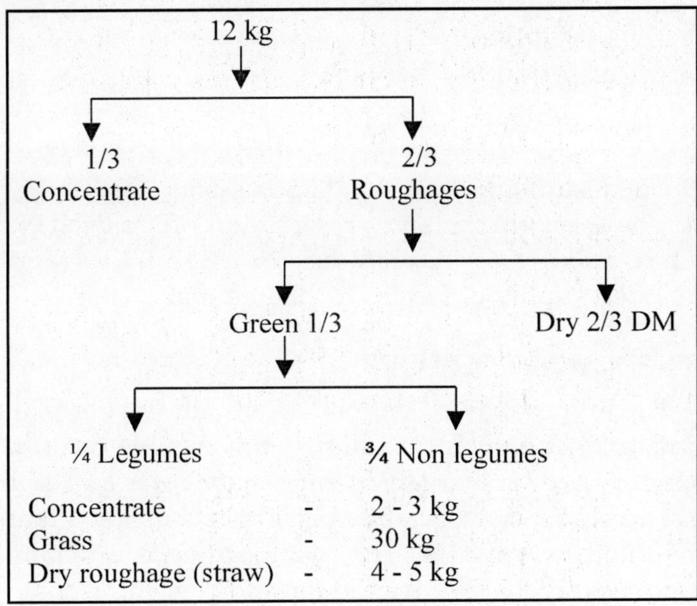

The area under fodder cultivation remained at the level of 4.4% of the cultivated area for the past 2-3 decades (80% grain crops, 15.6% other crops).

In order to improve the quality and quantity of forages in India, the forage scientists have been able to identify location specific fodder varieties and species and suitable cultivation technology for economic production. Forage production technology includes the development of different fodder production systems to suit wet, garden and dryland farming systems in addition to forage production in marginal and wastelands including problem soils such as saline, alkaline, acidic, ameliorated lands *etc*. The essential components of the system for maximizing fodder production include:

i) Improved forage varieties for increased fodder production per unit area per unit time
ii) Mixed cropping with legumes to improve fodder quality
iii) Silvipasture
iv) Tree leaf fodder
v) Non conventional fodder

The technologies on forage processing, fortification of low grade roughages and preservation of seasonal excess by hay making or by ensiling are essential to regulate continued supply of fodder for the animal.

2. FEED AND FORAGES

Feed types may be defined as any product of plant or animal origin, which the animals are capable of utilizing its organic and/or inorganic components. Forage in its broad sense refers to the vegetative portion of the plants that are consumed by animals. The common feeds and forages include green fodder, dry fodder and concentrates.

2.1 Classification of feeds

Livestock feeds can be generally divided in to two classes *viz.,* roughages and concentrates.

Roughages are bulky feeds containing relatively large amount of less digestible nutrients. Roughages and concentrates may therefore be characterized by three criteria *viz.*, dry matter, level of nutrients per unit volume/ weight and crude fibre content as shown in the table below.

Table 2. Dry matter and nutrient content

Forage group	Dry matter content (%)	Level of nutrient /unit weight	Crude fibre content (%)
Roughage	< 60	Low	> 18
Concentrates	> 60	High	< 18

2.1.1 Roughages

Roughages usually have low weight per unit volume and are often very coarse with low readily available carbohydrates. Most roughages also have a high content of cell wall material, with appreciable amount of lignin, cellulose, hemi cellulose, pectin, polyuronides and silica. Lignin and silica are negatively related to fiber digestibility i.e., with increasing lignin or silica content in the fodder, the fibre digestibility decreases.

Roughages are again divided into two types' viz., succulent (green) and dry, based on the moisture content of the forage.

Succulent feeds usually contain 60-90 per cent of moisture which include pasture, cultivated fodder weeds, bunt grasses, tree leaves, silage, root crops, etc. Dry roughage include hays, straws and bhusa. The green forage are either grazed by the animal or fed as soilage or green chop or as silage which are important for animal health and productivity as they contain considerable amount of minerals and vitamins essential for the cattle health.

2.1.2 Concentrate

Concentrate feed materials contain relatively high total digestible nutritients (TDN > 60 %) and less crude fibre (less than 18 %).They usually supply primary nutrients (protein, carbohydrates and fat) at a higher level and contain lower moisture. In general, concentrates contain high nitrogen free extract and total digestible nutrient. On the basis of crude protein (CP) concentrates are classified as energy rich concentrates when CP is less than 18% and protein rich concentrate when CP > 18 %. Based on the source of origin they are also grouped as feed of plant origin and feed of animal origin.

Byproducts obtained in the industries from processing of raw materials for the production of primary products. The byproducts obtained from grain processing, oil seed processing and pulse processing are the major and important feed ingredients governing more than 70 % of the total concentrate feed produced and utilized in the livestock industry. The annual availability of agro-industrial byproducts and crop residue in India is estimated as 409.7 mt.

i) Energy rich concentrates

Energy rich concentrates include grains, seeds, milling byproducts, molasses, roots and tubers. The common grains used in cattle feed include maize, barley, oats, sorghum and bajra. The milling byproducts commonly used to feed cattle include bran, flour, germ, gluten, grain screenings, grout, hull, malt sprouts and meals.

Molasses are concentrated water solution of sugar, hemi cellulose and minerals, obtained usually as byproducts of various manufacturing operation of the juices or extracts of selected plant materials. Cane or blackstrap molasses, beet molasses, citrus molasses and wood molasses are common example of this group.

The root and tubers concentrates include turnips, sugarbeets, carrots, swedges, mangolds and cassava tuber.

ii) Protein rich concentrates

Protein rich concentrates contain > 18 % CP. They may be plant proteins, animal proteins, non-protein nitrogen compound or single cell proteins. The most commonly used plant protein sources include oil seed meal, mustard cake, cotton seed cake, coconut meal, sesame meal, soybean meal, pulse protein and brewer's grains.

The animal protein sources include meat or meal or meal scrap, meat and bone meal, blood meal, feather meal, hatchery byproducts meal and fishmeal.

The common non-protein nitrogen feed stuff includes ammonia, amide, amines, amino acids and a number of ammoniated byproducts. The urea treated products such as urea mixed concentrates, urea molasses liquid products, urea mixed silage and urea in salt blocks are dominated ammoniated byproducts.

Single cell proteins are obtained from yeast, bacteria or algae. They have been specially grown or obtained as by products from various sources for use in animal feeds.

2.1.3 Commercial feed

A number of branded feeds are available in the market. The manufactured feeds may be divided into two major groups *viz.*, primary feeds and secondary feeds. The primary feed is one, which is mixed with individual ingredients, sometime with the addition of premix, at rates less than 40 kg t^{-1} of the finished feed. The secondary feed is he one, which is mixed with one or more ingredients and is a formula feed supplement.

The purpose of feed manufacturing is to provide feed that meets the intended specifications of nutritional composition, palatability desired, medication level and free of contaminants. The mixing process is the heart of any feed manufacturing system, which assures complete, thorough and uniform mix of all ingredients. The characteristics of quality are numerous within the context of manufacturing system. The quality begins with a balanced ration prepared by qualified nutritionist that meets the target species' nutritional requirement. The following are the characteristics of a good feed,

- The feed must be fairly consistent in its nutrient composition
- A most balanced source of nutrient
- Highly digestible
- Higher bio-availability of nutrients
- Agreeable taste
- Higher energy or protein content per kg of feed
- Higher feed conversion ratio
- Preferably a dense diet
- Least cost formulation

2.2 Forage classification

2.2.1 Grass fodder

The annual or perennial grasses are either cultivated or grown in pasture which are the major source for cattle Bajra Napier hybrid, guinea grass, para grass, anjan grass, setaria grass, blue panic, marvel grass, Rhodes grass, spear grass, deenanath grass etc., are the common grass species.

2.2.2 Cereal fodder

The cereal crops commonly cultivated for fodder purpose include maize, sorghum, pearl millet, teosinte, oats and fodder type small millets such as fox tail millet, barnyard millet and finger millet.

2.2.3 Legume fodder

The common legume fodder crops either cultivated for fodder or grown in the pasture include cowpea, lucerne, berseem, soybean, desmanthus, soybean, desmodium, cluster bean and stylosanthes. They are rich in protein, few of them are annuals, but many of them are perennials.

2.2.4 Tree fodder

A number of tree leaves are valued for their fodder and are much useful during summer or lean periods, when other green foliage are very much limiting. They are also called as top feeds e.g. subabul, sesbania, acacia and agathi.

2.2.5 Miscellaneous

This group includes sugarcane tops, weeds, mixed forages, bund grass, tuber crop leaves, etc. The green forages that are cut and fed in fresh condition to the animals are termed as soiling of crops.

Silage can be prepared during monsoon periods using excess fodder available from maize, sorghum, bajra napier hybrid grass, grass-legume mixture, legume and used in summer when the fodder availability will be very scarce.

2.3 Varietal features of forage crops

Table 3. Some of the cultivated fodder species for different regions of India

	Rainfed	**Irrigated**
(a) Arid tracts	Jowar, bajra, moth, guar, lobia	Lucerne, berseem, oats, maize, jowar, bajra
(b) Semi-dry	Bajra, jowar, lobia, moth, guar, velvet bean, field bean, moong	Jowar, maize, lobia, teosinte, lucerne, berseem, sarson, turnips, Hybrid Napier, oats, Sudan gras, guinea grass, *Setaria sphacelata,* Rhodes
(c) Semi-wet	Dinanath grass, jowar, lobia, rice bean, velvet bean, teosinte, sunnhemp	Berseem, oats, Sudan grass, Hybrid Napier, guar, jowar, maize, para, Rhodes, setaria
(d) Wet regions	Jowar, Dinanath, rice bean	Berseem, oats, Hybrid Napier, guinea, lucerne, berseem, sarson, turnips, Hybrid Napier, oats, setaria, para, jowar
(c) Lower hills	Jowar, lobia, bajra, velvet bean, field bean, guar	Maize, jowar, oats, berseem, lucerne, Hybrid Napier, Sudan, setaria, Rhodes

Some of the important high-yielding varieties of fodder grasses, cereal fodders and legumes, popular in different regions, are enumerated below.

Table 4. Important high-yielding varieties of fodder grasses, cereal fodders and legume

Crop	Place	Varieties
Jowar (*Sorghum vulgare*)	**Northern region** Punjab, Haryana, Delhi	J.S.-20, J.S.-29/1, J.S.-263, J3/53, Swarna, M.P. chari, S.L.-44, Pusa chari, Haryana chari
	Uttar Pradesh	T-3, T-4, 8B, M.P. chari','S-700, 'H_1', 'H_2', 'Rio'
	Maharashtra	Red Khaki, Nilwa,Nandyal, M-35-1
	Central region Madhya Pradesh	Gwalior-82, Gwalior-304, Vidisha 60-1, Ujjain-6, Ujjain-8, J-195
	Southern region Andhra Pradesh and Tamil Nadu	K 3, Co 11,Co 18, Co 19, Rungu 1, M.P. chari
Maize *(Zea mays*)	Andhra Pradesh, Karnataka and Tamil Nadu	Hybrids--Ganga-Safed-2, Ganga-3,'Ganga-5, Ganga-7, Composites--Jawahar, Amber, Kissan, Vijay, Sona, Vikram, Open-pollinated--N.P. Yellow-2, K-41, Bassi, Jaunpur
Bajra (*Pennisetum typhoides*)	Andhra Pradesh, Karnataka and Tamil Nadu	A-1/3, H.B.3, S-530, T-55D 1941,'2291
Oat (*Avena sativa*)	Andhra Pradesh, Karnataka and Tamil Nadu	Early varieties--Western-11,Mid-season varieties--Kent, Craig, Afterlee, Green Mountain, A-17, Fulgham, ,IGFRI-Soil-3021,IGFRI- 2688 Late varieties--Algerian, , FOS-1/29, Kharsai
Cowpea (*Vigna sinensis*)	Haryana, Punjab and Delhi	FOS-1, FOS-10, K-395, K-585, EC 4216, IGFRI-S-45, IGFRI-S-457
	Uttar Pradesh and Karnataka	Russian Giant, IGFRI-S-97, IGFRI-S-985, Russian Giant
	Gujarat	E.C. 4216, Russian Giant, Chhrodi 14-20, Chhrodi 26
	Southern states	Co 1, Russian Giant, E.C. 421628
	West Bengal	Co 1, E.C. 4216
Guar *(Cyamopsis tetragonoloba*)	Punjab, Haryana, Delhi, Uttar Pradesh	FOS 1, FOS 2, E.C. 4216 F.S. 277, IGFRI-S 212
	Rajasthan	2Durgapura safed
Velvet bean (*Stizolobium niveum*)	Uttar Pradesh, Rajasthan, Andhra Pradesh, Madhya Pradesh and Haryana	IGFRI-S 2276-5

Contd...

Table 4. Contd...

Field bean (*Dolichos lablab var. lignosus*)	All India	IGFRI-S 2214-II IGFRI-S 2218-1
Berseem (*Trifolium alexandrinum*)	All India	Meskawi-Diploid, Pusa giant-Tetraploid, IGFRI-S-29-1, Chhindwara
Lucerne (Medicago sativa)	All India	Type-8, 'ype-9 ,Anand II, IGFRI-S-244, Moopa, IGFRI-S-54, N.D.R.I.-1
Hybrid Napier	All India	Pusa Giant Napier, NB 2, EB 4, or Gajraj, N.B. 5, Coimbatore
Sudan grass (*Sorghum sudanese*)	All India	SS-59-3,G 287, Piper, J-69
Dinanath grass (*Pennisetum pedicellatum*)	All India	Type-3, 10, 15, or'IGFRI-S-3808, G-73-1, T-12, IGFRI-S-866-1
Anjan or Buffel (*Cenchrus ciliaris'*)	All India	Pusa Giant Anjan, IGFRI-S-3108, IGFRI-S-3133, C-357, C-358,
Marvel (*Dichanthium annulatum*)	All India	M-8,IGFRI-S-495-1,IGFRI-S-495-5

3. DETAILS ON FORAGE CROPS

3.1 Grasses

Forage, in its broad sense, include all feed consumed by the livestock. More generally, forage refers only to such material as pasturage, browse, hay, straw and silage. It is the vegetarian food for the domestic animals. Grasses, legumes, grasslands and pasture herbages grown specially for feeding the livestock, to be used either grazing or cut and fed, are also called as forages.

The principal forages are grasses in the family Poaceae and legumes in the family Fabaceae. There are more than 10,000 species of grasses grouped into 620 genera. The grasses are herbaceous plants, either annuals or perennials. They have variable habits *viz.* erect, prostrate, procumbent or decumbent. They are persistent in nature and have sod forming characteristics. They have immortal vigour and aggression. As bulk feed to the livestock, grasses are the best and cheapest.

Table 5. Forage crops and varieties for different situation in Tamil Nadu

Irrigated	Varieties
Perennial grasses	
Bajra Napier hybrids	Co-1, Co-2 and Co-3
Guinea grass	Co-1 and Co-2
Annual/biennial grass	
Deenanath	Co-1
Teosinte	
Cereal fodder	
Sorghum	Co-27
Bajra	Co-8
Maize	African Tall
Perennial pulses	
Lucerne	Co-1
Desmanthus	
Annual pulse	
Cowpea	Co-5
Beans, Cluster bean, Soybean	
Rainfed	
Perennial grasses	
Blue buffel	Co-1
Marvel grass, Rhodes grass, Spear grass	
Setaria grass, Signal grass, Sudan grass	
Congo signal grass	
Perennial pluses	
Desmanthus, Stylosanthes, Desmodium, Siratro, Clitoria, Centro, Rice bean,	
Velvet bean, Moth bean	
Cereal fodder	

The use of green grass as feed to livestock ensures normal health and reproduction of the animals. The major portion of the green fodder requirement of livestock is made up of grass. Grass is obtained from natural pastures and cultivated grass fields. Natural pastures serve as the grazing area for a vast majority of the livestock nearly, 41.57% of the total geographical area is the presently available grazing area for the animals in Tamil Nadu. The total grazing area is made up of (i) forests and forest outskirts (15.63%) (ii) Permanent pastures (1.15) (iii) barren and uncultivated wastelands (4.37%), (iv) cultivable wastelands (2.30%), (v) current fallows (12.0%), (vi) other fallows (4.70%) and (vii) miscellaneous trees and groves (1.42%). These areas are sometimes being diverted for roads, playgrounds, houses etc. which result in shrinking of grazing area year by year. Compared with the national average of 22.06% under forest and 3.92% under permanent pastures, Tamil Nadu has only 15.06% under forests and 1.15% under permanent pastures with the implications of increased forage production through cultivation, so as to meet the fodder requirement of the livestock.

As regards, the spread of the grazing are excluding the forests, it has been estimated to occupy an extent of 11.5 lakh ha in southern zone (Pudukottai, Chidambaram, Thirunelveli, Ramanathapuram, Virudhunagar) 7.39 lakh ha in NorthEastern zone (Chingleput, North Arcot, Villupuram, South Arcot and Thiruvannamalai districts) 5.78 lakh ha in Western Zone (Coimbatore and Erode districts), 3.08 lakhs ha in Cauvery Delta Zone (Thiruchirapalli, Thanjavur and Nagapattinam districts) 0.27 lakh ha in high rainfall Zone (Kanyakumari district). The Cauvery delta and southeren zones have in addition strips of coastal saline alluvium running approximately 1000km which also serve as grazing areas.

Different kinds of native grasses such as *Apluda arstata* (Mandapill), *Bothrichola intermedia, B. perstusa, Brachiaria ramose* (arisipillu), *Cenchrus ciliaris* (Kolukattai grass) *C. setigerus, Cynodon dactylon* (arugampillu), *C. plectostachyon* (Gaint star grass), *Dicanthium annulatum* (Marvel grass), *Elyonurus hirsutus, E. tripsacoides* (Sewan grass), *Eragrostis curvula* (Weeping love grass) *Hetropogon contortus* (Spear grass), *Ischaemum indicum, I. rugosum* (Kudukken pillu), *Iseileme laxum* (Tehengai narpul), *Panicum repens* (Injipul),

Pennisetum orientale, Sehima nercosum (Vennai pullu) etc. are the most grasses that are found to grow themselves under rainfed conditions in the natural pastures and grazing area.

3.1.1. Description of Forage Grasses

Bajra Napier Hybrid

Among grasses, Bajra Napier hybrids are the highest green forage yielder in unit time and space. Bajra Napier hybrid is an inter-specific hybrid between bajra and napier. The hybrid is a triploid and hence sterile. It is propagated by rooted slips or stem cuttings. It possesses high fodder yielding ability and perennial habit of napier and soft leaf and quality characters of bajra. It is more vigorous, nutritious, succulent, palatable and responsive to nitrogenous fertilizer than napier grass. The quality is good with high calcium, magnesium and phosphorus and very low oxalate content (2.8-3.2%). It is a tall growing (200-300 cm) erect, stout, deep rooted perennial hybrid grass. It is native of Rhodesia and South Africa. It is widely distributed in tropical and sub tropical regions of Asia, Africa, Southern Europe and America. In India, it is grown in Punjab, UP, Haryana, Gujarat, MP, Bihar, Orissa and West Bengal.

Guinea grass (*Panicum maximum*)

Guinea grass is a native of tropical and sub tropical Africa. It was introduced to India in 1793 in some of the military farms. It is an excellent fodder and pasture species and much valued for its high productivity, palatable herbage and good persistence. It grows well on a wide range of well-drained soils but not on heavy clay soils or in areas liable to prolonged water logging or flooding. It can be cultivated as irrigated crop as well as rainfed crop in areas with well distributed rainfall. It is shade tolerant and it is suitable for growing in coconut groves. It thrives between temperature ranges of 15-38°C.

It is a perennial grass and withstands multicut. It is propagated through seeds and slips. Seed dormancy has been recorded up to 5 months from harvest. Its nutritive value is quite high when leafy and young, but falls off with increasing maturity. Its protein content ranges from 5-8% and relished by all classes of livestock. The K, Ca, Mg contents in herbage is about 1.23, 0.52 and 0.69% respectively. The nutritive value mainly depends on the stage of herbage growth at harvest.

Para grass (*Brachiaria mutica*)

It comes up well under water undulated condition and with sewage water. It is otherwise called as water grass or buffalo grass. It is native of Africa and tropical South America and it is introduced to India in the year 1894. It is now widely distributed in humid tropical and sub tropical areas. It grows well on moist soil and withstands prolonged flooding or water logging.

It can be fed to cattle as green grass or hay. It hardly withstands heavy grazing. With heavy fertilization and sufficient soil moisture it gives up to 250 tha^{-1} green fodder yield. It can be cut at monthly interval. Under ordinary condition it may be cut at 6-8 week interval. It is a poor seed producer and generally propagated by slips.

Other *Brachiaria sp* such as *B.brizantha* (signal grass), *B. decumbens* and *B.ruziziensis* (Congo signal) are also high yielding forage species. Signal grass is semi erect, drought resistant, and perennial in nature with sharp, pointed deep green leaves about 35 cm long. The leaves remain green for longer time than other grasses. It thrives on loamy soil and is propagated by seeds and rooted slips. All the signal grasses come up well in sewage water. It contains 6 % protein and low value of 0.67 % Ca and 0.49 % P. Only tender shoot will be relished by the cattle and rejection will be high if any matured grass is fed.

Kolukattai grass (*Cenchrus sp*)

It is one of the dominant grass species of Indian, locally known as Dhaman grass and Anjan grass in India, Buffel grass in Australia and African Foxtail in United states and Kenya. The genus Cenchrus includes three important forage species viz. *C.ciliaris* (white kolukattai) *C.setigerus* (black kolukattai), *C.glacus* (Neela kolukattai). These three species are commonly grown in grazing lands of Tamil Nadu. White kolukattai is the most common pasture in the kangeyam tracts where the well known breeds of draught animals are bred.

White kolukattai (Buffel grass or anjan grass) is a native of North tropical and South Africa and Indonesia. The grass possess a wide range of adaptability. It is concentrated in the hot and dry tracts with an annual rainfall ranging from 250 to 750 mm. It withstand temperature upto 48^{0}C. It can grow in sub humid conditions upto an annual rainfall of 100 to 1280 mm. The grass has very high degree of drought resistance but affected by frost. The optimum temperature for its growth is 30^{0}C.

It grows on a wide range of soil. It establishes well on dry sandy soils in arid lands and sandy to sandy loam in semi arid regions where the soil is alkaline. The grass can also grow on alluvial, laterite and red soils. The grass thrives on calcium rich soils. Heavy clay, water logged condition and acid soils are not suited for its growth. The light textured soils of high phosphorus status are best for this grass. The optimum soil pH for its growth is 7 to 8 but it can grow on the soil with a pH as low as 5.5. The plants are rhizomatous in nature and are highly drought resistant. During summer months the plant remains dormant and with the receipt of rains they regenerate and produce enough canopies to cover the soil. It is a good soil binder and used in soil conservation bunds as a cover crop. It thrives exceedingly well in calcareous soil.

Black kolukattai (Birds wood grass) is similar to white kolukattai in its growth and adaptation but it is less dormant than white kolukattai grass in natural population. The two types of grasses are capable of producing 20-30 t ha^{-1} green fodder/year. They withstand heavy grazing and trampling by cattle. *Cenchrus glaucus* (Blou buffel) is popularly called as Neella kolukattai in view of the bluish tinch on leaf surface.

Spear grass (*Heteropogan contortus)*

Spear grass is an indigenous, perennial forest grass found to grow all over India in the semi dry areas. In Tamil Nadu, an extensive area of this grass is found in Salem and neighbouring districts, dominating and even suppressing other grasses especially on poor, rocky soils. It has the ability to grow on poor soils but thrives best on sandy loams having pH between 5.0 to 6.0. The species is fairly drought and frost tolerant but not to high level of salinity. It occurs in the areas receiving annual rainfall range between 500 to 1500 mm. It is hardy and quick growing upto a height of 60-120cm. It is to be cut and fed to cattle before flowering.

It spreads naturally very well. Pastures dominating with *Heteropogan contortus* are usually not fertilized. The grass is rarely sown and existing pastures are utilized for grazing. Otherwise, the sharp awns in the ear head make it unpalatable. The grass makes good hay and silage. It has a crude protein content of 8.75%, calcium 0.48% and phosphorus 0.16 %. It yields 20-25 t $ha^{-1}year^{-1}$. It is suited for rotational grazing at monthly intervals after proper establishment.

Marvel grass (*Dicantium annulatum*)

It is also known as karad. It is a tuffed perennial with stem 1m high, erect or geniculately ascending to form broad, some spreading tufts. Internodes are short and nodes are purple and hairy. Its roots reach up to 120 cm soil depth. It grows in areas receiving rainfall of 300-1500 mm and on variety of soils. It tolerates salinity but not acidity. In India, it is considered as best pasture grass. It is suited to medium to heavy soils. The grass does not thrive well in sandy soils. It has the capacity to withstand drought under arid and semiarid conditions of the desert and at the same time can bear water logging under rainfed or irrigated conditions. It is drought resistant perennial grass with bluish green leaves having a distinct yellowish midrib. It is common in Southern and Eastern districts of Tamil Nadu. It is largely used for hay making. It gives green fodder yield of 30-35 t ha^{-1} $year^{-1}$. It has a crude protein content of 3.1%, crude fibre content of 35.2%, calcium 0.44%, phosphorus 0.41% and potassium 1.0%.

Rhodes grass (*Chloris gayana*)

This leafy perennial grass is named after Cecil Rhodes. Origin is from Tropical Africa. It can grow both in Tropical

and sub tropical regions. It is drought resistant grass and can tolerate a fair degree of salinity. It can be established by seed as well as rooted cuttings. The seed sown crop will be ready for harvest in three months time while the vegetatively raised crop can be harvested in two months. Subsequently cuts are taken at monthly intervals. It yields about of 37 t $ha^{-1}year^{-1}$ with irrigation and 17 t ha^{-1} $year^{-1}$ under rainfed conditions. On account of its quick growing aggressive habit, the grass tends to get too crowded and hence, it is necessary to open up the earth once in a year during the rainy season. The grass is leafy and palatable and fairly nutritious with 5% protein, 0.5% calcium and 0.4% phosphorus.

Blue panic (*Panicum antidotale*)

A tall glabrous perennial grass with erect stems thickened at joints, plants grows upto the height of 2.0 to 2.5 meters. The leaves are fairly long, 35 to 60 cm in length ad about 12 m in width with acuminate tips and blue green leaves. The plant is very deeply rooted giving good drought tolerance and is adapted to the drier margins at 400-700 mm rainfall. It is a vigorous, summer grower, capable of shooting quickly in response to rainfall. It is burnt by frost but survives through to the next growing period. This is drought tolerant grass well suited for arid and semi arid regions. The plant is deep rooted and thin stemmed, perennial in nature, with smooth and solid culm and thickened nodes.

The herbage contains 7-8 per cent crude protein. The grass is less palatable as compared to *Cenchrus*. Its palatability was estimated to be 11 to 13 per cent of that of *Dicanthium*, one of the most palatable Indian grass. Its rapid growth produces large quantity of palatable herbage. The grass can withstand heavy grazing. Its stem rapidly becomes hard and woody and injures animal mouth. Grazing is restricted to current year's growth only. The grass should be cut or grazed before flowering because at late flowering stage it accumulate oxalic acid which is responsible for the cause of kidney disorders. It is seed propagated and the first cut can be taken four months after sowing and subsequently once in a month. It gives 20-25 t $ha^{-1}year^{-1}$ in three cuts under rainfed conditions. With sewage irrigation, it is capable of producing 150 t $ha^{-1}year^{-1}$. It is recommended as a cut fodder and also as a pasture. It contains 7.3% crude protein, 40.5% crude fibre, 0.39% Calcium, 0.09% phosphorus and 1.62 % potassium.

Deenanath grass (*Pennisetum pedicellatum*)

This is very popular in Africa and Australia as a fodder grass and was introduced into India in 1950's. It is an annual plant, erect in habit and profusely tillering. Under irrigated condition, two to three ratoon crops can be taken. It gives a green fodder yield of 40 – 50 t ha^{-1} in 55 - 60 days. It can be directly fed to cattle or ensilaged. It contains 18.2 % dry matter, 13.9 % crude protein, 3.1 % crude fat, 25.5 % crude fibre, 0.72 % calcium and 0.25 % phosphorus. It has a digestibility of 55.9 % (IVDMD).

Sudan grass (*Sorghum vulgare var sudanensis*)

It is introduced to India in the year 1920. It is a quick growing, drought resistant crop. It thrives well under warm climate in semi arid, arid, tropical and subtropical regions. It is propagated by seed. The grasses are harvested after completely flowering both for main and ratoon crops. In about four cuttings 45 –50 t ha^{-1} yield is obtained under irrigated condition. It is suitable for hay making but not for silage.

3.1.2. Cultivation practices of forage grasses

Field preparation

For both irrigated and rainfed grasses, especially perennial grasses, deep ploughing with disc/mould board plough is essential because forage grasses are maintained in the field for many years. Their fibrous roots must have more soil depth for effective foraging of soil moisture and nutrients. Deep ploughing is essential to eliminate unwanted weeds besides improving the infiltration of the soil for better rainwater harvesting.

Seeds and sowing

The seed of Bajra napier hybrid grasses are sterile. Hence, vegetative propagation either through root slips or stem cutting is done. For sole crop: 40,000 slips or stem cuttings ha^{-1} and for inter cropping with one row of Desmanthus @ 30,000 slips ha^{-1}.

One disadvantage of this is need of more volume of bulky stem cuttings/root slips which warrants additional expenses on cutting and transport. For guinea grass, seeds are available but the seed dormancy is for about 6-8 months and the germination is also very poor (20-30 per cent). Considering this aspect, propagation through root slips is better. In case of *Cenchrus* and other grass species, depending on the availability either root slips or seed are used. Almost all the grass seeds have dormancy at least for 4-6 months.

Season is important consideration for the success of the crop. Irrigated grass can be sown /planted throughout the year, since there is water availability.

Manures and fertilizers

Since all the grass species have fibrous root system, they get their nutrient mainly from the topsoil layer (0-30 cm depth). Grasses are perennial and they need sustained support through soil nutrient and improved physical condition of the soil. This can be achieved through adequate supply of FYM / compost @ 25 t/ha. Moreover, grasses are the most neglected crop among the cultivated crops. Hence, it is obvious that only marginal or poor lands are allotted to them. Under this situation not only application of FYM/ compost is essential, but also application of inorganic fertilizer such as NPK as basal is essential. For bajra napier hybrid, 50:50:40 of NPK kg ha^{-1} of blanket fertilizer application is recommendation. Apply full dose of NPK before planting by opening furrow 5cm deep on the side of the ridges and cover and 75 kg ha^{-1} of N after each cut. Application of NPK has been found to increase the fodder yield of grasses nearly two to four times. Biofertilizers such as Azosprillium + Phosphobacterium mixture can be used for both seed treatment (3 pockets or 600 g/ha) and soil application (10 pockets or 200 g/ha) before sowing. Biofertilizer is found to increase the fodder yield by 1-2 t/ha.

Weeding

Important operation to be done both for irrigated and rainfed grass at early stage of the crop, especially between 20-40 days after sowing/planting. Otherwise, the grass will be suppressed by the prolific growth of weeds and this may result in gaps in field (due to mortality of grasses) and ultimately results in reduction in the green fodder yield. Preferably weeding with hand hoes is better since it incidentally creates better aeration for soil by the way of scrapping the soil surface. Depending on the weed intensity, nature of grass growth, either one weeding (between 20-30 days) or two weeding (20th and 40th days) may be needed.

Under irrigated condition, weeding and earthing up is essential after every harvest. In rainfed condition, unwanted bushes, shrubs and prominent weeds should be removed as and when needed.

Irrigation

For irrigated grasses (CN grass, guinea grass, buffalo grass, Teosinte and deenanath grass), life irrigation should be given on 3rd day of sowing/planting. Thereafter depending on the soil type and climatic condition, irrigation must be given once in 7-10 days. If there is moisture stress due to some unexpected reason it will reflect on the yield.

Under rainfed condition compartmental bunding can be followed for better harvest of rainwater. If there is excess moisture/water, it has to be drained out. Water logging for more than 3 days may affect the growth of certain grasses such as *Cenchrus sp., Panicum sp.* etc.

Plant protection

It is not generally needed for bajra napier hybrid grass, guinea grass and buffalo grasses. But in teosinte and deenanath grass, there may be some pests (aphids shoot fly) and diseases (rust, leaf spot). Since they are mainly used as fodder crop, the occurrence of pests and disease at later stages (beyond 40 days) may not affect the crop yield and quality to a great level. If they occur in the early stage, plant protection measures have to be taken. The crop should not be fed to animals for about 30 days to avoid toxicity.

Harvest

For bajra napier hybrid grass, guinea and buffalo grass, first harvest is due on 75-90 days. Subsequent harvests are done at 45 days interval. Totally there will be 7-8 cuts year^{-1}. Deenanath grass and teosinte are harvested on obtaining 50% flowering, which roughly coincides with 60-65th day after sowing. Delay in harvest may reduce the quality of fodder in terms of less crude protein and more crude fibre. Other rainfed grasses can be harvested at 50% flowering. Depending on the rainfall, 2-4 cuts year^{-1} may be obtained from rainfed grasses.

Yield

The green fodder yield ranges from 300-350 t ha^{-1} in respect of bajra napier hybrid grass. For guinea grass it ranges from 200-250t ha^{-1} and for buffalo grass it ranges from 120-180 t ha^{-1}. Deenanath will yields about 50-60 t ha^{-1} and teosinte yields about 30-35 t ha^{-1}. Kolukkattai grass yields about 20-30 t ha^{-1}. The other rainfed grasses yields on an average 10-15 t ha^{-1}year^{1}.

Table 6. Choice of grasses for different situation

No	Systems and Name of the grasses	Varieties
I	***Irrigated***	
Ia	***Perennials***	
	Bajra Napier Hybrid grass (*Pennisetum glaucum X P.purpureum*)	Pusa Giant, IGFRI – 3, 6, 7 & 10, NB -21, BN - 2, Co1 , Co2 , Co3
	Guinea grass (*Panicum maximum*)	Co1, Co2 , PGG – 1, 9, 13 & 14
	Water grass or Buffalo grass (*Brachiaria mutica*)	No improved varieties
Ib	***Annuals/Biennials***	
	Deenanath (*Pennisetum pedicellatum*)	Co1, PS- 3, IGFR - 3808
	Teosinte (*Zea mexicana*) Syn. (*Euchlena mexicana*)	Improved Teosinte, TL - 1
II	***Ranifed (perennials)***	
	Blue buffel (*Cenchrus glaucus*)	Co1
	White kolukattai or Anjan grass (*Cenchrus ciliaris*)	IGFRI 3108,Marwar, Anjan (CAZRI 75)
	Black kolukattai grass/Dhaman grass (*Cenchrus setigerus*)	Marwar, Dhaman(CAZRI 76)
	Marvel grass (*Dicanthium annulatum*)	GMG - 1
	Rhodes grass (*Chloris gayana*)	

Contd...

Table 6. Contd...

Spear grass (*Heteropogen contortus*)	
Setaria grass (*Setaris anceps*) and *Setaria sphacelata*	Nandi, Kazungula, Narok ,PSS 1
Blue panic/Australian grass (*Panicum antidotale*)	
Congo signal grass (*Brachiaria ruziziensis*)	
Brachiaria grasses Signal grass (*Brachiaria decumbens*) *Brachiaria brizantha*	
Arisi pillu (*Brachiaria ramosa*)	
Sudan/Johnson grass (*Sorghum sudanese*) (*Sorghum halepense*)	
For hilly areas (Temperate)	
Kikuya grass (*Pennisetum clandestinum*)	
Tuber grass (*Phalaris tuberosa*)	
Rye grass (*Lolium multiflorum, L.perennae*)	
Dactilis grass (*Dactilis glomerata*)	
Setaria grass (*Setaria anceps; S. sphacelata*)_	

3.2 Cereals

3.2.1 Description of cereal fodder crops

Maize (*Zea mays*)

Maize is one of the most important economic plant of the world, serving as food for man and forage for cattle. Its origin is Mexico.

Maize is an almost ideal forage crop, as it is quick growing, high yielding, palatable and nutritious.

It can be cut for fodder within two months after sowing and unlike sorghum, there is no risk of prussic acid poisoning by feeding young plants. Although it is not risky, it is not so hardy or so widely adaptable to soil and climatic conditions as sorghum. The yields from maize remain fairly uniform through out the year.

Sorghum (*Sorghum bicolor*)

Next to wheat and rice, sorghum is the most important cereal crop of the world. It has been in cultivation since very ancient times in all regions of the tropics where rainfall is low and uncertain. Sorghum is almost unique in its ability to grow over a very wide range of climatic and soil conditions. Sorghum is believed to be indigenous to Africa. The name sorghum is derived from the '*Sorgo*', meaning 'rising above' to denote its tallness as compared to other cereal crops. It is the largest as well as the most important amongst the group of cereals known as millets and hence known as the 'Great Millet', Since it is well adapted to arid and semiarid conditions of the world.

Pearlmillet (*Pennisetum glaucum*)

Bajra is mainly raised as grain crop, but can also serve as quick growing fodder crop under irrigation, which is nearly equal to sorghum green fodder. Quick growing, emergency fodder, its better tillering facilitates 2-3 cuttings within 75 days and absence of HCN poisoning. Unlike sorghum, there is no risk of prussic acid poisoning and bajra can be cut and fed even before flowering.

Finger millet (*Elusine corocana*)

Finger millet is not cultivated as fodder crop. Locally known as ragi. Finger millet straw is nutritious fodder and

can be fed green or as hay. Straw is harvested as green and fed to the animals. Ensiling of ragi straw enhances its feeding value. The straw from dry and rainfed areas can be converted to hay, and can be used for feeding of cattle largely in Karnataka. It tolerates salinity than other cereal crops. Ragi straw from irrigated crops is tough and fibrous and makes poor fodder, but improves on stacking, as fermentation occurs and makes it less tough and palatable. Rainfed ragi straw is however, of good quality and it is readily consumed by the cattle.

Oats (*Avena sativa*)

Oats rank next to wheat and barley. Indian oats are mainly cultivated in Punjab and Uttar pradesh. The plant is erect annual, with lax, profuse panicle and long glumes. It can be cultivated in temperate regions and cool sub tropical paces. It can be grown on all soil types, expect on very sandy or alkaline soils.

Teosinte (*Zea mexicana*)

It is native to central America and its one species *Euchalaena mexicana*, has been introduced to India as a fodder crop. This plant is wild relative of maize. It is tall, leafy, succulent, annual, growing to a height of 1.8-3.6 m in large clumps with numerous basal branching tillers.

The plants are monoceious, like maize. In Punjab it is recommended for growing as a green fodder in the fodder scarce months of May - June and October - November.

Foxtail millet (*Setaria italica*)

It is a rapid growing short duration annual cereal growing to a height of 80-110 cm, may be cut for hay or green soiling within 50 days of sowing. It is occasionally grown as fodder crop. It has wider adaptability in tropics and is well suited for most soils. It is not good for making hay and silage. Marginal and sub marginal lands can support a good crop of fodder. It has high protein content and well relished by the cattle.

Varagu *(Paspalum scorbiculatum)*

It is widely distributed in the tropics. It grows to a height of about 60-90 cm and has a considerable long duration, taking about 80-85 days for 50 per cent flowering. Widely distributed in tropics and cultivated for grain and fodder.

Panivaragu (*Panicum miliaceum*)

It is very short duration crop and number of days for 50 per cent flowering range between 34-38 days. The plant is rich in protein (11.93-13.92% crude protein).

Samai (*Panicum miliare*)

It grows to a height of 65 cm and comes to 50 per cent flowering between 43-50 days of sowing. On an average it gives a green fodder yield of 10-12 t ha^{-1}.

Panivaragu and samai are sometimes grown for fodder purpose. They grow more quickly than even thenai and are considered as emergency fodder. They are highly suitable for cultivation in hot weather under light irrigation or in garden land as rainy crop under cold weather season.

3.2.2 Cultivation practice of cereal fodder crops

Fodder sorghum (*Sorghum bicolor*)

Season

1. Irrigated: Jan to Feb and April to May
2. Rainfed: June - July: South West monsoon
   Sept - Oct: North East monsoon

Varieties

Irrigated: CO 11, CO 27

Rainfed: K7, CO 27

Soil: All soils, but loamy soils with good drainage are best suited.

Field preparation: Plough once with Iron plough and twice with country plough for rainfed crop. Field should be prepared in advance taking advantage of early showers.

Apply FYM @ 12.5 t ha^{-1} before the crop and incorporated

Apply 10 packets of azospirillum (2kg ha^{-1}) for irrigated crop.

Forming ridges and furrows for irrigated crop

1. Form ridges and furrows of 6 m long and 30 cm apart
2. Form irrigation channels across the furrows using a ridge
3. If ridges and furrows are not made form beds of size $20m^2$ depending on the availability of water.

Application of fertilizer

For irrigated crop: 60:40:20 kg NPK ha^{-1}

Basal: 30: 40: 20 N P K kg ha^{-1}

Top dressing: 30 kg N ha^{-1} on 30 DAS

For rainfed crop: 40:20:0 kg ha^{-1}. Entire dose as basal

Seed rate

Irrigated crop: 40 kg ha^{-1}

Rainfed crop: 75 kg ha^{-1}

Treat the seeds with 3 pockets of Azospirillum (600 g ha^{-1})

Spacing: 30 x 15 cm for both the crops.

Irrigated crop: Sow the seeds to a depth of 3 cm and cover the seeds

Dry Crop: Use seed drills for sowing at 5 cm depth or use country (pre monsoon sowing) plough and sowing behind the country plough.

Water management for irrigated crop: Irrigate immediately after sowing, life irrigation on 3rd day and there after once in 10 days.

Weed management: Along with hand weeding, thin the seedling and use it for gap filling and maintain the spacing of 15 cm between plants.

First weeding: 20 DAS

Second weeding: 30-40 DAS if necessary.

Harvesting: If it is a single cut, it should be harvested at 60 to 65 days (50% flowering) and if it is a multicut variety, the first cut is at 60 DAS and a second cut 40 days after Ist cut.

Yield: First cut: 45 t ha^{-1} (green)

Second cut: 25 t ha^{-1} (green).

Fodder maize (*Zea mays*)

Season: Irrigated through out the year.

Rainfed: Kharif / Rabi season.

Field preparation: Plough once with Iron plough and twice with country plough for rainfed crop. Field should be prepared in advance taking advantage of early showers.

Soil: All soils with good drainage.

Manures and Manuring: FYM/compost: 12.5 t ha^{-1}

Fertilizer recommendation: 60: 40: 20 kg NPK ha^{-1}

Basal 30: 40: 20 kg N P K ha^{-1}

Top dressing: 30 kg N on 30 DAS

Irrigation

Form ridges and furrows using ridger, 6m long and 30 cm apart / form beds depending on the availability of water.

Varieties: African Tall, Ganga - 5

Seed and Sowing

Maintain seed rate of 40 kg ha^{-1} and dibble one seed to spacing of 15 cm between the seeds in the row to a depth of 4 cm.

Spacing: 30 x 15 cm

Seed treatment : Treat the seeds with 3 pockets (600 g) Azospirillum inoculant before sowing.

Weed management : Hand weeding on 20th day and subsequent weeding if necessary.

Water management : Irrigate immediately after sowing and life irrigation on 3rd day and thereafter, once in 10 days.

Harvesting: Harvest the crop when the cob is in milking stage (50% flowering to dough stage).

Green fodder yield: 40 to 45 t ha^{-1}

It can be mixed with cowpea.

Fodder bajra / Bajra (*Pennisetum glaucum*)

Seed rate: 10 kg ha^{-1}

Spacing: 30 X 10 cm

Manures and Manuring

FYM: 12.5 t ha^{-1}, NPK: 50:40:20 kg ha^{-1}.

25:40:20 kg ha^{-1}applied as basal prior to sowing, the remaining 25 kg N as top dressing on 25 DAS.

Seed and Sowing: Sow the seeds to a depth of 2 to 3 cm and cover / broad cast and cover with country plough.

Variety: Co 8

Hand weeding: 20 DAS and subsequent as and when required.

Irrigation: At sowing followed by life irrigation on 3 DAS and subsequently once in 10 to15 days interval

Harvest: First cut 40 to 45 after sowing at boot leaf stage.

Yield: 30 to 35 t ha^{-1} cut^{-1}. 3 to 5 cuts can be taken in CO 8. It can be intercropped with fodder cowpea to improve the fodder quality.

Teosinte (*Euchlaena mexicana*)

Climate

Being a tropical species can be grown in warm humid regions with annual rainfall of > 1000mm.

Soil

It needs rich well drained loamy soil for best growth.

1

Season

Usually grown in kharif season. Best time for sowing in North India is June 25th to July15^{th}. It is also grown during Rabi season.

Sowing

Seed rate: 40 kg ha^{-1}. Either broadcast or drill sowing, adopting 30cm row spacing.

Manures and Manuring

12.5 t ha^{-1} or FYM. NPK at 20:40:40 kg ha^{-1}. Entire NPK as basal before sowing.

After cultivation

One weeding is sufficient. It requires higher amount of water.

Harvest

110 to 120 days after sowing. Sometimes second cut 40 to 50 days after first cut.

Fodder yield: 40 to 50 t ha^{-1}.

3.3 Legumes

The word legume is derived from the Latin word '*legere*' which means to gather because the pods have to be gathered or picked by hand as distinct from reaping the cereal crop *viz.,* poaceae species. Man used legumes as early as 5000 BC in the Bronze age by the Lake Dwellers of Switzerland, in 3000 BC by Chinese and in 3000 – 2000 BC by Egyptians and later on by Americans, Indians and others. Legume plants produce and supply the major portion of the world's plant protein. This is utilized directly by man through eating the seed or grain, or indirectly by livestock feeding and subsequent consumption of livestock products. Fodder legumes also referred as masals and have immense value in animal nutrition, because of their higher protein content (19-24%), vitamin specific minerals like P, Calcium etc. and rich in crude protein content. Legume forages are nearly equal to concentrates and are likely to be substituted for the latter.

Legumes can be grown as pure crops separately, cut and fed along with grasses in mixtures or can be grown with grasses or cereal fodders in a right proportion to get high yields coupled with high quality.

A number of varieties in different species of the below mentioned genus have been developed and recommended for cultivation as nutritious forage legumes. Some of them are Medicago (Alfalfa or Lucerne), Vigna (Cowpea), Cyamopsis (Guar or Seeniavari / Cluster bean), Phaseolus (Mothbean or Naripayiru, Phasy bean), Trifolium (Berseem, Egyptian clover, Persian clover), Desmanthus (Velimasal), Stylosanthes (Muyal masal), Trigonella (Mehta or vendayam), Melilotus (Senji Pilipesara, or sweet clover), Lab-lab (Avarai/Motchai), Glycine (Soybean), Clitoria (Sangupu), Macroptilium (Siratro), Sesbania (Sithagathi), etc. Of the several crops of leguminous forage, a handful like Lucerne, cowpea, naripayir, desmanthes and stylosanthes are popularly cultivated by farmers in Tamil Nadu. Some of them are grown under irrigated conditions. Naripayir, stylosanthes, muyal masal and clitoria are best suited for rainfed situations. A brief note on some of the important leguminous forage crops in given hereunder.

3.3.1 Importance of forage legumes

1. Short duration in nature
2. Raised as catch crop in between two crops
3. Improves soil fertility by way of 'N' fixation
4. Suitable for inter or mixed cropping
5. Dual purpose. e.g., Sunhemp, Berseem
6. Increase intake of fodder by improving fodder availability
7. Rich in protein, calcium, phosphorous and minerals
8. Capable of replacing concentrates in animal rations and save feeding costs.

3.3.2 Description of legume forage crops

Lucerne / Kudirai masal (*Medicago sativa* L.)

Lucerne is called the "queen of forage crops". It is known as alfalfa, which in Arabic language means 'best fodder'. In Tamil Nadu it is known as kudirai masal. It has its origin in South - Western Asia but has spread all over the world, as it is well adapted to warm temperature and cool sub-tropical regions. In India, it was introduced in the year 1900.

It is a perennial legume and once established as an irrigated crop, it continues to give fodder for four years. Legumes are well adapted to extremes of heat and cold in a dry region but it cannot tolerate even moderately high temperature in moist climate. Soil acidity and greater incidence of diseases in humid areas hinder its successful cultivation. It can be cultivated on a variety of soils from sandy loam to clay, but it grows best on well-drained deep loam soils of good fertility. Highly alkaline soils are not suitable for its cultivation. It requires a seed rate of 20 kg ha^{-1} with a fertilizer recommendation of 24:120:40 kg ha^{-1}

Kandhar or Quetta, Persian or Arabian and Meerut were the varieties under cultivation in earlier period. LLC-9, LLC-14. Anand-3 and Co-1 are currently under cultivation in different parts of the country. The best time for sowing lucerne is from the middle of October to the middle of November. On light soils, the seeds may be broadcasted in a moist seedbed at the rate of 20 kg per hectare. On heavy soils, sowing on ridges 45 to 60 cm apart is commonly practiced. In this case, the seed is dibbled by hand 1.5 to 2.5 cm deep on the top of the ridge and the land is immediately irrigated, taking care that the water does not swamp the ridge tops. Growing on ridges economises the irrigation water. Sowing of seed inoculated with Azospirillum is beneficial. The first cutting is generally low and is taken 2½ to 3 months after sowing. The subsequent cuttings are taken at intervals of 3 to 4 weeks depending upon the growth and when the plants are in flowers. Eight to ten cuttings can be taken in a year, giving a total green fodder yield of 80 to 100 t ha^{-1}.

Lucerne is highly nutritious and palatable. It is specially used for feeding horses, poultry and draught animals. It contains 20 per cent crude protein, 2.3 per cent calcium and 0.23 per cent phosphorous. Milk cattle fed exclusively on lucerne suffer from bloating due to the content of saponin glucoside and to avoid this, green lucerne should be chaffed and mixed with an equal quantity of dry bhusa. Feeding of lucerne at the flowering stage and dried for sometime lessen the harmful effect. Animals fed on lucerne hay seldom suffer from bloating, cattle may be fed with one third of lucerne admixed with two third of grasses. Upto 10 kg of lucerne can be given to an animal in a day.

For seed production the crop is left uncut after February and irrigated sparingly during the flowering period. The seed crop is harvested in April or May. Under average conditions, yields about 200 – 250 kg of seed per hectare.

Dodder (Cuscuta) is a problem in Lucerne. It is a total stem parasite and gains entry into the field only through the crop seed as admixture collected from an already infested field. Cuscuta seeds germinate along with lucerne and infest the crop resulting in poor yield. The cheapest and simplest method of eradicating cuscuta is to remove the parasite along with the host and burn it. Lucerne seeds should be free of contamination from cuscuta. Cultivation of lucerne in fields once infested with cuscuta should be avoided.

Hedge Lucerne – Velimasal (*Desmanthus virgatus*)

Hedge Lucerne is a recent introduction from Thailand. Since the time of introduction in the year 1976 in Tamil Nadu, intensive research efforts were made and it was

identified as a valuable and highly nutritious leguminous fodder. It belongs to the sub-family Mimoceae under the family leguminacea. It is a perennial shrub producing many upright branches bearing bipinnate compound leaves made of minute leaflets. It withstands repeated cuttings and regenerates quickly.

It is suitable for cultivation in all types of soils and seasons alike. It comes up well under irrigated conditions but performs equally well under unirrigated conditions endowed with drought tolerance. It contains 19.2 per cent crude protein and 27 per cent dry matter with 55.3 per cent Invitro Drymatter Digestability (IVDMD). It has no toxic principles and can be cut and fed to animals at any stage of its growth. It requires a seed rate of 20 kg ha^{-1} with a fertilizer recommendation of 10:60:30 kg ha^{-1}. Hedge lucerne called as Velimasal in Tamil, grows to a height of 2 m in about 60 days after its establishment. It can be cut at 1m height and the lopping used for feeding the livestock.

Hedge Lucerne can be raised as a pure crop adopting a spacing of 45 cm between rows. Seeds at the rate of 15 kg per hectare are hand-drilled in rows spaced at 45 cm. It is recommended for cultivation as an intercrop in tall growing grasses like bajra napier hybrids and guinea grass. It is highly compatible to be grown with the grass. One row of hedge lucerne for every three rows of grasses is an ideal combination and a quantity of 225 t ha^{-1} of balanced feed of legume and grass can be obtained. As a pure crop hedge lucerne yields about 125 t of green fodder per hectare.

Stylosanthes-Muyal masal (*Stylosanthes scabra*)

It is an exotic species recently introduced and acclimatized to South Indian conditions, comes up well

under rainfed condition. It is a hardy type and extremely drought resistant. It is a perennial herb, attaining a height of 60 to 75 cm in about six months time. It produces many semi-erect branches with lots of secondary and tertiary branches assuming a bushy appearance. It has very slow growth in the initial period of its establishment.

Once established, it grows fast and puts good canopy cover made of tripinnately compound leaves. First cut can be made in about six months and subsequent cuts at every three months afterwards under rainfed conditions. If grown under irrigated conditions, it can be cut at every 45 days at a height of 45 cm. It has got quick regeneration capacity. It yields up to 30 to 35 t ha^{-1} of green fodder as pure crop. It has a protein content of 19 per cent.

Muyal masal is best recommended as an intercrop in the rainfed Blou buffel or Kolukattai grass. Grown in one row for every three rows of grasses to get a balanced and nutritious fodder. When the crop growth is optimum, both muyal masal and kolukattai grass can be harvested, slightly dried in the field and preserved as hay.

Sangupushpam (*Clitoria ternatea*)

It is an indigenous species, perennial in nature and very well adapted to rainfed conditions. It is an indeterminate, winy creeper and fast in its growth. There are two distinct types, one white coloured and the other blue coloured petals. White flowered type is high yielding than blue flowered type.

First cut can be made in 60 days after sowing and subsequent cuts can be taken at every 45 days. It can be grown all along the fence or as a border crop. It contains 18 per cent protein.

Desmodium (*Desmodium tortuosum and D. unsineatum)*

Desmodium is a nutritious and highly palatable legume fodder and is called "alfalfas of the tropics" or tropical medic. Two tropical species *Desmodium tortuosum* (Green leaf desmodium) and *D.unsineatum* (Silver leaf desmodium) are being popularized in Tamil Nadu. Green leaf desmodium is an erect perennial, growing to a height of 2 to 2.5 m and is distinguished by the brown flecking on the upper surface of hairy leaves. It comes up well in acid soils. Silver leaf desmodium has an irregular silver band along the margin of the leaf.

Livestock relish both the leaves and young shoots. It is an irrigated forage legume and perennial in habit. Desmodium is a shade tolerant forage legume and comes up under shades of coconut, tamarind and eucalyptus trees. It can be cut at 30 cm height for four times a year. The crop under shade yields 25 t ha^{-1} of green fodder. It contains 21 per cent protein and 27 per cent dry matter. It is a self sown crop. Seeds shattering in the ground germinate at optimum soil and moisture conditions. Desmodium pods are in a chain like form and maturation proceeds from top downwards. It is as good as alfalfa. It is used as protein supplement in poultry feeding.

Cowpea (*Vigna unguiculata*)

Central Africa is considered to be its native home though it is claimed indigenous to India. It can be grown

throughout the year yet it is best suited for kharif season. It grows well on practically all types of soil but well drained moderately textured soil is good for its growth. It requires a seed rate of 40 kg ha^{-1} with a fertilizer recommendation of 25:40:20 kg ha^{-1}. The crop is ready in about 55 to 60 days and should be cut when the first pods are about to ripen. A pure crop yields about 22 - 25 t ha^{-1} of green fodder and about half this quantity under rainfed conditions. The value of the protein rich fodder lies in the fact that it is grown in mixture with cereal fodders like cholam, maize or bajra to make the fodder more nutritious. It is grown in alternate rows with maize or after every two rows of cholam or bajra.

Co-5 fodder cowpea is an ideal fodder type for Tamil Nadu. It is quick growing, profusely branching and twining in habit, it yields about 35 t ha^{-1} of green fodder in 60 days. HFC 42-1 is another quick growing variety with non-twining habits. It has good regeneration capacity and is amenable for multi-cut, provided first cut is taken at 15 cm height after 50 days of sowing.

Cluster bean (*Cyamopsis psoraloides*)

Commonly called guar or seeni avarai or kothavarai is an important drought resistant annual fodder as well as a soil restorative crop, especially suitable for light sandy or alluvial soils. It cannot withstand excessive moisture or water logging. It is grown for vegetable and fodder, and seed for gum purposes. The seeds are sown by broadcast or hand drilled in rows 30 to 45 cm apart adopting a seed rate of 25 to 40 kg per hectare. As forage crop, it is often sown mixed with fodder jowar to provide a balanced cattle feed.

The crop becomes ready for harvest in about 60 to 75 days and yields about 16 to 20 t of green fodder per hectare under irrigation and half of this under rainfed condition. Though guar is not yet cultivated for fodder purposes in Tamil Nadu, fodder varieties like FS 277, HFG 119 and HFG 156 are quite popular in Northern India. Guar yields about 28 to 35 t ha^{-1} of green fodder. It contains 16 per cent protein.

Moth bean - Naripayir (*Phaseolus aconitifolius*)

It is the most drought resistant among the kharif pulses and is largely grown as dry crop either alone or subsidiary to millets or as a component in crop mixtures. On account of its spreading habit, it is highly useful against wind erosion in sandy areas. It can be used both as green and dry fodder. Seeds are broadcasted at the rate of 10 to 15 kg per hectare when grown alone or 5 to 10 kg per hectare when mixed with other crops, particularly Kolukkattai grass or cholam. It is a low yielder and a green fodder yield of 10 t per hectare can be obtained.

Lab-lab-Field bean / Avarai (*Lab lab niger*)

Primarily, grown as a pulse and it is grown for fodder also. It is quick growing, profusely branching and twining in habit. It is grown as an inter crop with rainfed cholam in kharif season. It yields about 30 t ha^{-1} of green fodder in 65-70 days. It is nutritious and highly palatable.

Soybean (*Glycine max)*

It is a native of sub-tropical central Asia and is one of the important crops in China, Japan and other Mongolian countries. It can be grown alone or mixed with sorghum or maize. It is particularly valuable, as it keeps green during October - November when other green fodder is scarce. It will grow on all types of soil, the best result being obtained on fertile loam and sandy loam soils. It is largely cultivated in North India and it is gaining importance as a pulse (grain) and fodder crop in Tamil Nadu.

Both erect and creeping varieties are available, the later being more suitable for fodder purpose. Pusa selections yellow and chocolate are creeping types with good fodder yield. Monelta is a short duration variety.

The seed is sown in rows 45 to 60 cm apart at the rate of 25 to 30 kg per hectare. The crop yields 13 to 16 t of green fodder per hectare in about 60 to 70 days. The fodder is palatable and rich in protein and other nutrients.

Berseem or Egyptian clover : (*Trifolium alexandrinum*)

It is indigenous to Egypt and introduced to India (1904). In India, it is an important rabi forage crop in Punjab, Haryana, Delhi, Rajasthan, Gujarat, UP etc. It is a very good fodder for milch animals and horses. Climate requirement for the cultivation of berseem is dry and cool climate. It contains 20% crude protein, 70% dry matter digestibility and rich in Ca and P. It is also used as green manure for improving physical, chemical and biological properties of soil and also decreases the soil bulk density and results in better soil aggregation. It cannot with stand drought and frost. It cannot be grown in damp and heavy rainfall areas. All type of soils is suitable for cultivation except very sandy soils. Well-drained medium loam soils rich in P and calcium. It performs well in acid soil. It requires a seed rate of 20 to 25 kg ha^{-1}. Blanket recommendation of NPK is 25:60:0 kg ha^{-1}.

Temperate forage legumes

Different species of clover, such as red clover (*Trifolium pratense*), white clover (*T. repens*), crimson clover (*T.incarnatum*), aslike clover (*T. hybridum*), and subterranean clover (*T.subterraneum*), centrosema (*Centrosema pubsecens*), calapogonium (*Calapogonium mucunoides*), vetches (*Vicia sp*) etc., are the important forage legumes suitable for cold temperate places and high elevations. These are usually grown in mixtures with grasses and maintained as perennials. A quantity of 8 to 10 kg of seed is required for sowing with grasses. For pure crop 40 to 50 kg of seeds are required. These contain 15 to 16 per cent crude protein. All of them are resistant to frost and yield 100 to 150 quintals of green fodder per hectare. Centrosema and calapogonium are suitable for growing under shades of plantation crops such as tea, coffee, coconut etc.

3.3.3 Cultivation practices of forage legume

Field preparation

Two ploughings either with country plough or with tillers (tractor drawn) is needed. For irrigated crop either beds and channels or ridges and furrows are formed. In rainfed condition, only two ploughing are done and one more shallow ploughing is given to cover the seeds, especially after sowing.

Seeds and sowing

In irrigated condition seed rate is generally less because line sowing is done with proper spacing (30-40 X 10 cm). Whereas, in rainfed sowing seed rate is generally high because sowing is done by broadcasting method.

Manures and fertilizers

FYM at 25 t ha^{-1} for irrigated crop and at 12.5 t ha^{-1} for rainfed crop is applied. For irrigated crop both basal (30:40

kg NP ha^{-1}) and top dressing (20 kg N ha^{-1}) is done. For rainfed crop only basal dressing is done with a fertilizer dose of (20:40 kg NP ha^{-1}). If the rainfall is favorable, top dressing is resorted to 10-20 kg N/ha. Biofertilizers *viz.,* Azospirillum + Phosphobacterium mixture both as seed treatment and soil application is recommended @ 3 packets (600 g) for seed treatment and 10 packets (2000 g) for soil application per hectare.

Weeding

One or two weedings before 40th day is required to have a good stand of crop. If weeding is not done, weeds will suppress the crop and ultimately the green fodder yield gets reduced. Weed control by using herbicide is not done usually, fearing residual toxicity on the animals.

Irrigation

Life irrigation is given on 3rd day of sowing for irrigated crop. Subsequent irrigation is given once in 7-10 days depending on the soil and climatic conditions. Totally about 6-7 irrigations are needed.

Mixed cropping

For improving the quality of fodder, legumes are generally mixed with cereals both under irrigated and rainfed conditions. Trails conducted at TNAU, Coimbatore revealed that legume as intercrop with cereal fodder increased the crude protein content.

Table 7. Crude protein content of fodder as influenced by crop mixtures

Crop mixture	Crude protein content (%)
Sorghum + Legume	13.94
Bajra + Legume	14.98
Maize +Legume	14.58
CN H grass + Legume	12.82

Cropping system approach for year round fodder supply

Since most of the cereal fodder crops are shorter in duration (60-75 days), they are amenable to fit in a cropping system. Experiments conducted in Tamil Nadu revealed that the following cropping sequences in a year were found to give more green fodder yield.

1. Maize + Cowpea - Maize + Cowpea - Maize + Cowpea = 106 t ha^{-1}
2. Sorghum + Cowpea - Maize + Cowpea - Maize + Cowpea = 110 t ha^{-1}
3. Bajra Napier Hybrid grass + Hedge Lucerne (Desmanthus) = 225 t ha^{-1}

Plant protection

Generally plant protection is not needed. However, some pests such as shoot fly, stem borer, aphids etc., in sorghum, maize and bajra. Similarly diseases such as downy mildew, leaf spot etc., may occur. If the pests and diseases occur in the early stage of the crop (before 30th day), plant protection should be taken by the application of pesticides or fungicides and care should be to use the fodder after 30 days of application.

Harvest

First cut on 90th day after sowing at 50 cm height and subsequent cut once in 30-45 days at the same height. In case of rainfed crop, 4-5 harvest per year based on the growth, which mainly depends on soil moisture content.

Yield

Green fodder yield varies greatly with crop and variety.

3.4 Trees

The trees which provide leaves, flowers, stems and pods palatable to animals are called fodder trees. Multipurpose trees, shrubs and woody components have been traditionally included in the farming system management as a source of five F's *viz.*, fuel, fodder, fibre, food and fertilizer.

3.4.1 Importance of fodder trees

- Fodder trees provide nutritious top feed in the form of leaves and pods rich in proteins and minerals to livestock throughout the year and especially during the lean period.
- They provide shade to the grazing animals, which increase the efficiency of digestion of feed.
- They provide a variety of products such as fuel, small timber, fibre, human food, medicines etc.
- Sources of organic matter, increases the total level of soil nutrients besides improving the soil texture.
- The fodder trees serve as lining fences and hedges besides yielding green fodder.
- Serve as wind breaks besides intercepting the wind blown soil particles by the tree canopy and washing to the ground beneath.
- Provide shade for cocoa, coffee and tea plants and supports for yams, vanilla and black pepper.
- Prevent soil erosion and conserve moisture.
- Increase the yield and improve the quality of grasses grown under the shade due to change in the microclimate condition.

A wide range of tree species, both leguminous and non-leguminous has been identified as suitable for feeding to animals.

Table 8. Tree legumes commonly used in Asian farming system

Species	Common name	Ecological adaptation	Growth form and characteristics	Major uses	Other uses
Acacia nilotica	Babul	Dryzone, drought tolerant	Thorny deciduous	Fodder	Timber, Fire wood
Acacia tortolis	Israeli babul	Dry arid zone tolerant to drought alkalinity	Flat topped tree to 15m, thorny	Fodder	Fire wood
Albizia lebbeck		Wide adaptability dry humid tropics	Tree to 25m	Fodder	Timber, Fire wood
Cajanus cajan	Vagai	Wider adaptability	Annual to perennial	Fodder, green manure	Fire wood
Calliandra calothrysus		Humid tropical	Shrub to 8m fast growing acid tolerant	Fodder, green manure	Timber, Fire wood
Cassia siarrea		Semiarid to sum humid	Shrub to 8m, coppices vigorously, drought tolerant	Fodder	Fire wood
Dalbergia sissoo	Sissoo	Tropical and sub tropical midlands, sub humid to humid	Tree to 30m	Fodder, green manure	Timber, Fire wood, shade tree
Erythrina spp		Humid lowlands to highlands	Tree to 20m, often thorny coppices vigorously	Fodder green manure,	-
Gliricidia sepium		Humid to sub humid lowland tropics	Tree to 20m coppices vigorously	Fodder, green manure	Fire wood, shade tree
Hardwickia pinata	Aachal Anjan	Dry tropics, drought tolerant	Tree to 30m grow slowly	Fodder	Fire wood
Leuceana leucocephale	Subabul	Humid to sub humid	Tree to 20m coppices vigorously	Fodder, green manure	Fire wood, shade tree
Pithecelobium dulce	Pungam	Wide adapatablility from dry to humid zones	Tree to 20m often thorny	Fodder	Fire wood
Pongamia pinnata	Vanni	Low to mid and humid tropics	Small tree to 8m, salt tolerant	Fodder	Timber, Fire wood, shade tree
Prosopis juliflora	Khejri	Dry zones, heat and alkali resistant	Tree to 10m coppices vigorously	Fodder, green manure	Fire wood, shade tree

Contd...

Table 8. Contd...

Sammanea saman	Rain tree	High to lowland tropics	Tree to 20m coppices	Fodder	Timber
Sesbania grandiflora	Agathi	Low to midland humid tropics	Spreading tree to 40 m fast growth	Fodder, green manure	Fire wood
Tamarindus indica	Tamarind	Wider adaptability	Large tree to 30m.	Fodder	Fire wood, shade tree

3.4.2 Description of fodder tree crops

Subabul (*Leuceana leucocephala*)

Subabul is known as Ipil-Ipil, Soundal. Its growth rate is fast especially under irrigated conditions. Leucalna produces large biomass per unit area. The variety Co-1 released, as an improved cultivar is a selection from Giant Ipil-Ipil K-28.

It gives a green fodder yield of about 100 t ha^{-1} in 2 years from 13 cut. It has a dry matter content of 25%, crude protein of 26%, IVDMD of 46%, Iron 121 ppm, copper 10 ppm, manganese 29 ppm and mimosine 3% (lowest). The yield of fuel from leucaeana under rainfed condition is about 28 t ha^{-1} in 4 ½ years and under irrigated about 75 t ha^{-1}. For fodder it has to be cut once in 45 days if under irrigation and at 60 days interval if rainfed.

To reduce the mimosine content in subabul

1. Soak the leaf material in hot water approximately 60°C for 3 minutes.
2. Dry the leaf material at 70°C

It should be mixed with other grasses. e.g., 20-25% of the total green fodder as subabul. Any animal should not be fed exclusively with subabul.

Agathi (*Sesbania grandiflora*)

Under irrigated condition it gives 60 t ha^{-1} of green fodder yield and under rainfed condition 25 t ha^{-1}. It can be fed to animals once in 60 days. It contains 23% protein, 2% calcium and 0.3% phosphorous. It is rich in amino acids like arginine, histidine, isoleucine, leucine, methionine, phenylamine, thiomine and tryptophan. The flowers can also be fed and they contain 2% crude protein and vitamin C.

Sithagathi (*Sesbania sesban*)

This is a small growing to about 2-4 meters, profusely branched with small oblong leaves and yellow or dark pink flowers with bright spots. This is a fast growing drought resistant fodder tree. It has a crude protein content of 22% and 1.1% Ca in leaves. Immature pods can also be fed to animals.

Glyricidia sepium

It comes up well in any soil including acid soils and wastelands. First cut can be taken in one year and subsequently once in 4-6 months. It can also be grown as a fence giving a spacing of 45 cm between plants. Since the leaves have a smell, initially the animals may not consume the fodder. The leaves can be wilted for few hours and then can be fed or the animal may be starved for few hours and then fed with the leaves. Once the animal gets accustomed to the green leaves, it will continue to feed. None of the leguminous fodders should exceed 20-25% of the daily green fodder ration.

3.5 Forage cropping system

Enterprises are complementary with each other when the production of one increases the quantity of other from given resources. The complementary nature of grasses and legumes is perhaps the most important explaining their place in the farm organization.

The main beneficial effect of growing grass/legume + legume together and including fodder crop in cropping system are,

- Effective utilization of limited land resource to increase the forage production per unit area
- Enhancement of quality of forage crops
- Reduction in the fertilizer cost by intercropping legumes and grass in cropping system
- Preservation of the land for future
- Improvement in Physico-chemical and biological properties of soil under grass-legume mixture.

Interactions between component crops greatly alter their stand establishment, growth rhythm, morphological development and nutrient uptake in cereal-legume mixed or intercropping systems. These, in turn influence the forage yield and quality traits. The grasses and the cereal forage crops do not have enough protein to support the growth and production of animals. However, by associative cropping of graminaceous and leguminous forages, the nutritive value of mixed herbage can be greatly improved as legume component contains higher amounts of protein, calcium and phosphorous. This ensures proper mineral balances in animal ration, curtails the use of feed supplements and produces milk at cheaper rate.

In multiple forage cropping systems three or four appropriate forage crops are grown in quick succession in a year to enhance the herbage productivity per unit area. The degree of its success depends upon agro climatic conditions and management levels. On the basis of the experiments conducted under the All India Command-ordinated Research Project on Forage Crops, the promising intensive forage crop rotations have been identified for different agro-ecological zones of the country. It is observed that the northern, central and western zones are the most efficient with green forage production ranging from 168 to 255 t ha^{-1} $year^{-1}$. Eastern, southern and hill zones, on the other hand appear to be less productive, yielding green forage from 102 to 225 t ha^{-1} $year^{-1}$.

The supply of nutritious forage throughout the year could be regulated by staggered sowing and forage conservation strategy. In order to achieve desired level of quality fodder, proper tillage, timely sowing, phased harvesting and adequate inputs are essential.

Productivity of forage systems involving perennial and annual forages is encouraging in yield and quality fodder production. The development of an overlapping cropping system of berseem + mustard - hybrid napier + cowpea for supply of green forages round the year under 'cut and carry' programme is a proven technology for specialized dairy farms. With integrated input management, this system provided green forage at 286, 240 and 175 t ha^{-1} $year^{-1}$ in first, second and third year respectively. At a planting spacing of 1.0 x 0.5 m, the contribution of perennial grass and annual crops work out to be 56 and 44 per cent respectively.

This kind of production system combined with forage conservation techniques could provide feed resources sufficient to maintain 10 to12 high producing milch cows with little or no supplementation of concentrates.

In the above mentioned forage cropping system, hybrid napier could be replaced by relatively soft and palatable species like setaria or guinea grass. The ability of guinea grass to remain succulent and thus palatable with greater acceptability even under the situations of delayed harvesting.

These perennial grasses could be planted in paired rows of 60 –75 cm spacing to facilitate the use of farm machineries in cultural operation for berseem in winter and for cowpea in summer. Both berseem and cowpea can be replaced with lucerne, which grows well and provides cuttings throughout winter and summer months. The management strategies should include,

- Rejuvenation of grass stumps in the form of stubble shaving to maintain its perenniality and productivity.
- The fertilizer schedule should consists of 40 kg N ha^{-1} at each cut of grass (200 kg ha^{-1} for each five cuttings), in addition to annual basal application of 20-25 tonnes of FYM, 60 kg P_2O_5 and 40 kg K_2O ha^{-1}.

3.5.1 Advantages of forage systems

- Continuous flow of green forage from the same piece of land
- Efficient use of inter row space through intercropping of berseem during winter season when grass remains dormant
- Saving of water and fertilizer as grass is established in the standing crop of berseem and cowpea is grown on residual soil fertility.

- Protection of dormant grass from frost injury by berseem as live mulch
- An associated legume nourishes the growing grass during spring and summer and improves the herbage quality
- Improvement in physico-chemical and biological properties of the soil under grass legume mixture

3.5.2 Inter cropping

Intercropping perennial fodder grass with perennial legume at 3:1 ratio is recommended for higher protein harvest. Of the different combinations test verified with bajra napier hybrids, *Desmanthus* proved to be more ideal and the combined grass-legume mixed fodder contained 12 per cent crude protein which is almost 50 per cent higher than pure grass fodder.

A field trial was conducted at Coimbatore during summer season under irrigated condition in a clay loam soil. There were five treatments, *viz.,* sorghum alone (T_1), maize alone (T_2), Cowpea alone (T_3), sorghum + cowpea (T_4) and maize + cowpea (T_5). The experiment was laid out in randomized block design with four replications. There was a uniform basal application of 60 and 30 kg N and P_2O_5/ha, respectively. The cereals and legumes were mixed in the ratio of 60:40 respectively, and sown in lines 25 cm apart. The cowpea seeds were inoculated with rhizobial culture before sowing. The crop was cut after 65 days from the date of sowing and the yield of green fodder was recorded.

The crude protein content of cowpea registered the highest value of 18.4 per cent and the lowest value being observed in the pure crop of sorghum (7.7%). The protein content of the fodder increased when the cereals were grown in combination with legumes. With regard to protein yield on dry matter basis, maize + cowpea treatment was significantly higher (801 kg ha^{-1}) than sorghum alone (564 kg ha^{-1}) and maize alone (494 kg ha^{-1}).

The P and Ca contents were significantly higher in pure cowpea crop than cereal alone. The mineral content was appreciably improved when the cereals were grown along with legumes. Phosphorus had increased from 0.23 per cent in pure sorghum to 0.32 per cent in mixed crop and Ca from 0.72 to 1.20 per cent. The pure crop of maize had 0.27 per cent and 0.80 per cent of P and Ca, respectively, whereas, the cowpea plus maize mixture had registered 0.41 per cent and 1.32 per cent of P and Ca. It could be concluded that the protein and mineral contents of cereal fodder could be improved by growing them in association with legumes.

It was seen that sorghum grown with soybean in the ratio of 2:1 yielded on an average 45.9 t ha^{-1} of green fodder, 8.1 t ha^{-1} of dry matter and 1142 kg ha^{-1} of crude protein against 33.9 t ha^{-1} 6.9 t ha^{-1} and 616 kg ha^{-1}, respectively, of sorghum pure crop and 11.7 t ha^{-1}, 2.1 t ha^{-1} and 369 kg ha^{-1}, respectively, of soybean pure crop. The green fodder have been increased and improved in the mixed crop of sorghum and soybean, compared to the pure crops of sorghum alone or soybean alone.

A field experiment was conducted under rainfed conditions to study the effect of growing legumes as intercrop with grasses in improving the quality of fodder. Two grasses, *viz., Cenchrus* and Blue panic and two legumes, *viz., Clitoria* and *Stylosanthes* were raised as sole crop and as intercrop with a ratio of 3:1 (three rows of grass and one row of legume) consisting of four grass legume mixture combinations. The results revealed that growing grasses with legumes as inter crop increased the crude protein content, compared to growing grasses as sole crop. The increase in crude protein in both *Cenchrus* + *Clitoria* combination and *Cenchrus* + *Stylosanthes* was 41.3% compared to *Cenchrus* as sole crop. Similarly, the increase in crude protein content for Blue panic + *Clitoria* was 37.4 and 30 per cent for Blue panic + *Stylosanthes* as compared to Blue panic as sole crop.

Though the legumes had high crude protein content (11 to 16%), the crude protein yield was low (0.35-0.47 t ha^{-1}) as compared to crude protein yield of grasses (0.59-0.70 t ha^{-1}) because of their low dry matter production per unit area. However, in the grass : legume mixture combinations there is an increase in the total crude protein yield. The combination of *Cenchrus* + *Stylosanthes* seems to be the best showing the highest crude protein yield (0.97 t ha^{-1}) besides highest green fodder yield (40.9 t ha^{-1}) and dry matter yield (10.16 t/ha).

Intercropping of sorghum with cowpea/soybean was evaluated for green matter, dry matter and crude protein yield in different ratio. Sorghum + Soybean in ratio of 2:1 was found to record significantly better yield of green fodder (56 t ha^{-1}), dry matter (9.9 t ha^{-1}) and crude protein (1.5 t ha^{-1}) as compared to pure crop of sorghum (34.6, 7.0, 0.6 t ha^{-1}, respectively). The results indicated that growing of legumes (soybean) in association with cereal crops like sorghum is beneficial for obtaining high dry matter and crude protein yield from a unit area than growing them as pure crop.

4. FORAGE QUALITY

Forage quality can be defined as the extent to which a forage has the potential to produce a desired animal response. Factors that influence forage quality include the following.

Table 9. Fodder rotation in various regions of India

Rotations at various centres	Greenfodder yield (t ha^{-1})
(1) Jhansi	
1. Hybrid napier + cowpea - berseem + sarson	286.3
2. Maize + cowpea - M.P. chari – berseem + sarson	197.2
3. M.P. chari - turnips – oats	
(2) Hyderabad	
1. Hybrid napier + cowpea - Hybrid Napier + cowpea - Hybrid napier + Berseem	133.4
2. Maize + cowpea - bajra + cowpea + berseem	126.7
(3) Anad	
1. Hybrid Napier	287.7
2. Hybrid Napier + guar - lucerne	252.9
3. Maize + cowpea - maize - cowpea - oats - maize + cowpea	168.5
(4) Kalyani	
1. Maize + cowpea - *P. pedicellatum* - oats	130.8
2. Maize + cowpea - rice bean - berseem + sarson	111.5
3. Maize + cowpea + jowar + cowpea - oats	88.4
(5) Kanke	
1. Maize + cowpea - oats - bajra + cowpea	102.6
2. Jowar + cowpea - berseem + sarson – maize + cowpea	96.0
3. Bajra + cowpea - berseem + sarson – maize + cowpea	95.9
(6) Pantnagar	
1. Napier + berseem intercropped and cut at the optimum time	214.1

Contd...

Tabe 9. Contd...

2. Napier + berseem intercropped and cut at the same time	199.8
3. Napier + lucerne intercropped and cut at the optimum time	196.0
(7) Jorhat	
1. Hybrid Napier alone	144.2
2. Maize + cowpea - maize - jowar - oats	66.4
3. Guinea alone	60.7
(8) Hissar	
1. Napier - bajra hybrid intercropped with berseem	211.7
2. Napier - bajra hybrid + lucerne	176.0
3. Berseem + Japan rape - jowar + cowpea - jowar + cowpea	170.5
(9) Coimbatore	
1. Sorghum + cowpea - maize + cowpea - maize + cowpea	110.7
2. Maize + cowpea - maize + cowpea - maize + cowpea	106.0
3. Guinea grass round the year	93.5
(10) Palampur	
1. Maize + cowpea - lucerne + oats + sarson	84.4
2. Maize + cowpea - turnip - oats + pea - cowpea	83.3
3. M.P. chari + cowpea - oats + pea - cowpea	78.2

Contd...

(11) Jabalpur	
1. Hybrid Napier intercropped with cowpea - berseem and cowpea	176.1
2. M.P. chari - cowpea - berseem + sarson - jowar + cowpea	168.6

(Feed and forages , Krishi world)

4.1 Terms of forage quality

Palatability

Animals select one forage over another based on smell, feel and taste. Palatability may therefore be influenced by texture leafiness, fertilization, dung or urine patches moisture content, pest infestation or compounds that cause a forage to taste sweet, sour or salty. High-quality forages are generally highly palatable.

Intake

Animals must consume adequate quantities of forage to perform well. Typically, the higher the palatability and forage quality, the higher the intake.

Digestibility

Digestibility (the extent to which forage is absorbed as it passes through an animal's digestive tract) varies greatly. Immature, leafy plant tissues may be 80 to 90% digested, while less than 50% of mature, stemmy material is digested.

Nutrient content

Living forage plants usually contain 70 to 90% water. To standardize analyses, forage yield and nutrient content are usually expressed on a dry matter (DM) basis. Forage dry matter can be divided into two main categories: (1) cell contents (the non-structural parts of the plant tissue such as protein, sugar, and starch); and (2) structural components of the cell wall (cellulose, hemicellulose, and lignin).

Anti-quality factors

Various compounds may be present in forage that can lower animal performance, cause sickness, or even result in death. Such compounds include tannins, nitrates, alkaloids, cyanoglycosides, estrogens, and mycotoxins. The presence and/or severity of these elements depend on the plant species present (including weeds), time of year, environmental conditions and animal sensitivity. High-quality forages must not contain harmful levels of anti-quality components.

Animal performance

It is the ultimate test of forage quality, especially when forages are fed alone and free choice. Forage quality encompasses "nutritive value" (the potential for supplying nutrients, i.e., digestibility and nutrient content), how much animals will consume and any anti-quality factor present. Animal performance can be influenced by any of several factors associated with either the plants or the animals. Failure to give proper consideration to any of these factors may reduce animal performance level, which in turn reduces potential income.

4.2 Factors affecting forage quality

Many factors influence forage quality. The most important are forage species, stage of maturity at harvest, and (for stored forages) harvesting and storage methods. Secondary factors include soil fertility and fertilization, temperatures during forage growth, and variety.

Species differences

Legumes vs. grasses

Legumes generally produce higher quality forage than grasses. This is because legumes usually have less fiber and favour higher intake than grasses. One of the most significant benefits of growing legumes with grasses is improvement of forage quality.

Cool-season vs warm-season grasses

Forage grasses are divided into two broad categories: cool season (adapted to temperate regions) and warm season (best adapted to tropical or subtropical environments). Cool-season grasses include orchard grass, bluegrass, perennial and annual ryegrass. Bermudagrass, bahiagrass, dallisgrass and corn are examples of warm-season grasses. Cool-season species are generally higher in quality than warm-season grasses.The digestibility of cool season grass species averages about 9% higher than warm-season grasses. Minimum crude protein levels found in warm-season grasses are also lower than those found in cool-season grasses. Within each category, annual grasses are often higher in quality than perennials. Due to differences in leaf anatomy (tissue arrangement or structure), warm-season grasses convert sunlight into forage more efficiently than cool-season grasses, but their leaves contain a higher proportion of highly lignified, less digestible tissues.

Temperature

Plants grown at high temperatures generally produce lower quality forage than plants grown under cooler temperatures and cool-season species grow most during the cooler months of the year. However, forage of any species tends to be lower in quality if produced in a warm region rather than in a cool region.

Maturity stage

Maturity stage at harvest is the most important factor determining forage quality of a given species. Forage quality declines with advancing maturity. For example, cool season grasses often have dry matter (DM) digestibilities above 80% during the first 2 to 3 weeks after growth initiation in spring. Thereafter, digestibility declines by 1/3 to 1/2 percentage units per day until it reaches a level below 50%. Maturity at harvest also influences forage consumption by animals. As plants mature and become more fibrous, forage intake drops dramatically.

Leaf stem ratio

Reduced leaf-to-stem ratio is a major cause of the decline in forage quality with maturity, and also the loss in quality that occurs under adverse hay curing conditions. Leaves are higher in quality than stems and the proportion of leaves in forage declines as the plant matures. Reproductive growth lowers leaf-to-stem ratio and thus forage quality. Most cool season grasses require a period of cool temperatures (vernalization) for flowering, so they produce reproductive stems only in the spring. Thus, the forage quality of regrowth of these grasses is greater and changes less over time because they have higher leaf-to-stem ratios than first-growth forage. Legumes and some grasses such as bermudagrass can flower several times each season, so their forage quality patterns are less closely linked to season.

Grass-legume mixtures

Grass-legume mixtures generally have higher crude protein concentration and lower fiber concentration than pure grass stands.

Fertilization

Fertilization of grasses with nitrogen (N) often substantially increases yield and also generally increases crude protein levels in the forage. Fertilization usually has little or no effect on digestibility. Fertilization with phosphorus (P), potassium (K), or other nutrients that increase yield may actually slightly reduce forage quality when growth is rapid. Excessive levels of some elements such as potassium may in some cases decrease the availability of other elements such as magnesium (Mg) in the diet.

Daily fluctuations in forage quality

Plants accumulate soluble carbohydrates during daylight and then use them overnight. Thus, soluble sugars are lowest in the morning and highest after a day of bright sunshine.

Variety effects

In species such as timothy that have a wide range of maturity dates, late maturing varieties tend to be slightly lower in digestibility because early types make more of their growth under lower temperatures. Some silage corn varieties have higher grain content and/or stover digestibility than others. The development of multifoliate alfalfa varieties (having more than three leaflets per leaf) is a strategy aimed at increasing forage quality, but some multifoliate varieties have no higher leaf percentage than traditional trifoliate varieties. Some trifoliate varieties exhibit superior quality, but care should be taken to assure that a "high-quality" variety is not substantially lower in yield.

Harvesting and storage effects

Leaf shatter, plant respiration and leaching by rainfall during field drying of hay can significantly reduce forage quality, particularly with legumes. Quality losses also occur due to weathering, plant respiration, and microbial activity during storage. In high rainfall areas, losses can be large for round bales stored outside, due to weathering of the outer layers.

4.3 Anti- nutritional factors or toxic constituents in forages

Hydrocynic acid (HCN)

HCN poisoning causes sudden death with in 1-2 hours.

Etiology

It is found in Sorghum, Cynodon, Johnson grass and Sudan grass. Young leaves contain more than 500 ppm.

Critical level

Less than 20 mg HCN / 100 gram of feed material (< 200 ppm)

Management

- Harvest for fodder at heading stage at 50% flowering stage
- Drying or haymaking
- Ruminants are more susceptible than horse and pigs.

Nitrate poisoning

Forage that accumulates more than 1.5% of NO_3 on dry matter basis are classified as potentially toxic.

Causes

Salivation, teeth grinding, high pulse rate abdominal pain, difficult breathing and finally death of ruminants.

Etiology

Found in immature green oats, Hybrid cumbu napier grass, rye, Para grass etc.

Management practices

- Irrigation management to prevent long spell of drought

- Dilution of high NO_3 water and cattle shed washings
- Avoid 'N' fertilizer application particularly during drought period
- Delayed cutting - allow over maturity
- Application of FYM / Compost.

Oxalates

Etiology

Oxalic acid present in Napier grass and rice straw (contain 1.5 to 1.6%). Young leaves of bajra contain more oxalic acid than matured leaves. Toxic limit up to 3 to 7 %.

Causes

Causes negative 'Ca' balance

Precipitation of calcium in the blood

Impairs P, Mg and Na nutrition.

Management

- Ensiling or hay making the Napier grass
- Combine with legume fodder
- Supplemental with chalk or supernated lime water at 1.0 l / animal may be given along with drinking water.

Mimosine

Etiology

A toxic amino acid found in subabul and *Mimosa pudica*. Toxic level: < 0.75%.

Causes

Infertility, goiter, low birth weight and death of new born.

Management

Mixing with other cereal forages / feeding 1% $FeSo_4$ in the diet / supplemented with Iron.

5. FODDER PRESERVATION

The rainfall in India is seasonal, all of which is not properly utilized. We can also produce a good amount of fodder during this season with proper selection of the plant material and proper cultural practices. To satisfy the needs of the stock during the lean months, namely, November-December and April to June, an adequate amount of the surplus grass available during the rainy season must be conserved. The ideal and simple method of conversion is to drive off the moisture in the fresh grass with artificial heat and store the products as dried fodder for use when required. Unfortunately, considerable capital expenditure is involved. In practice, the moisture in grass is reduced through exposure to sun and wind and hay is obtained. This practice is simple but is brought with difficulties in practice.

To combine cheapness and simplicity, and to ensure at the same time a product of high feeding value and virtually independent of weather conditions at the time, natural fermentation must be used and the process of ensilage adopted. There lies the choice of the farmer for hay or silage making.

5.1 Hay Making

Hay refers to grasses or legumes that are harvested, dried and stored as 85-90 per cent dry matter. High quality hay is green in colour, leafy and pliable and free from mustiness, less moisture at the time of storing. Hay can be utilized as an excellent feed for dairy cattle, particularly when fodder is scarce or pasturage is insufficient.

5.1.1 Principles of hay making

The principle involved in hay making is to reduce the water content of the herbage so that it can safely be stored in mass without undergoing fermentation or becoming mouldy. This must be accomplished in such a manner that the hay is not leached by rain and that the loss of leaves is kept at a minimum. Good, well cured hay makes an excellent forage for all types of livestock on a dairy farm, particularly at times when good quality green fodder is not available or that the green fodder available is too succulent and feeding of dry fodder in addition is essential to keep the stomach of the animal replete.

5.1.2 Good quality hay

1) Good hay should be leafy. It has been found that leaves are richer in food value compared to other parts of the plant. The leaves are generally rich in protein, vitamins and minerals. Loss of leaves in hay making would mean deterioration in feeding value of the ultimate product.

2) It should be prepared out of herbage, cut at a stage nearing maturity, preferably at the flowering stage when it has the maximum of nutrients. Delay in cutting would mean losses of a part of nutrients, which would be used up by the plant in seed formation.

3) It should be green in colour. The green colour of leaves indicates the amount of carotene which is a precursor of vitamin A.

4) Soft and pliable

5) Free from dust and moulds

6) Free from weeds and stubbles

7) Have the smell and aroma characteristics of the crop from which it is made

8) The moisture percentage in hay should not exceed 15 per cent

9) Hay of average quality will usually run from 25-30 per cent crude fibre and 45-60 per cent total digestible nutrient

10) Hay is primarily a cattle, buffalo, horse, sheep and goat feed. Very little of hay of any kind is ever fed to swine.

5.1.3 Kinds of Hay

Legume hay

Good legume hay has many characteristics that make it special value to dairy cattle. It has a higher percentage of digestible nutrients. It has more of digestible proteins because of the high protein content. Furthermore, the protein of legumes is of superior quality as compared to proteins from other plants. Well-cured legume hays are higher in vitamin contents. They are particularly rich in carotene and may even contain vitamin D. They are also a rich source of vitamin E. The legume hays are particularly rich in calcium and are generally palatable. Among various leguminous fodder cops lucerne, berseem, cowpea and soybean hays are considered first.

Non-legume hay

Non-legume hays made from grasses are inferior to legume hays. They are, as a rule palatable and contain less protein, minerals and vitamins than legume hays. Non-legume hays have the advantage over legume hays because their out turn per hectare is more than that of legume hays and the former can be grown easily.

For good quality hay making crops should be harvested in the milk stage. They are low in proteins and minerals, but rich in carbohydrates.

Mixed hay

Hay prepared from mixed crops of legumes and non-legumes is known as mixed hay. The composition of such a kind of hay will depend on the different species grown as a mixed crop. Such a crop is generally cut earlier because of the variation in the seeding time of the mixed crops. If harvested early, cereals are generally richer in proteins.

Table 10. Composition of hays

S.No	Hays	Crude protein	Crude fibre	Nutrients-free extract
1	Grass hay	%	%	%
	Harali hay	6.8	24.2	56.0
	Marvel grass hay	5.2	38.5	44.5
	Rhodes grass hay	12.7	27.3	45.8
	Spear grass hay	5.6	38.2	46.0
2	Legume			
	Berseem hay	14.7	30.6	41.0
	Cowpea hay	25.2	13.8	43.6

5.2 Silage

Silage may be defined as the green succulent roughage preserved under controlled anaerobic fermentation in the absence of oxygen by compacting green chops in air and water tight receptacles. The siling process, in the absence of oxygen, leads to the fermentation of water soluble carbohydrates to produce organic acids which increases acidity of the material (pH about 4.0). Such anaerobic acid production (mainly lactic acid) arrests the growth of bacteria and moulds as well as inactivities putrefying organisms, ie., acts as a change in nutritive value. It serves as one of the best method of fodder conservation. Unlike hay making, which has seasonal constrains and heavy nutrient losses, silage on the other hand is superior and has better acceptability, digestibility and nutritive value.

5.2.1 Process of ensiling

The ensiling operation includes the following steps, harvesting the crops at the optimal stage of maturity, wilting (if needed and possible) to ensure adequate dry matter content for the solidstate lactic acid fermentation, chopping, loading into a silo, compaction and sealing to exclude air, storing and finally unloading for animal feeding. Additives can be applied at the chopping stage or when loading into the silo.

The biochemical and microbiological events during the ensiling fermentation can be divided into four distinct stages,

a. *Aerobic stage :* When air is still present between the plant particles and the pH is still high, 6.0-6.5. During this stage plant respiration continues, as well as proteolysis and the activity of aerobic microorganisms such as enterobacteria, fungi and yeasts.

b. *Fermentation stage :* which is carried out by a dynamic succession, which change according to the conditions prevailing in silage, starting with enterococci and leuconostoc, followed by lactobacilli and pediococci (Woolford, 1984). Lactic acid and other organic acids accumulate and the pH decreases, below 5.0, depending on plant composition and its buffering capacity.

c. *Storage stage :* when the silage is sealed and no air penetrates, few changes occur.

d. *Unloading stage :* for feeding when the silage is re-exposed to air which re-activates aerobic microorganisms, mainly yeasts and moulds, which might spoil the silage.

It is desirable to accelerate the fermentation step as far as possible in order to minimize the activity of detrimental microorganisms and stabilize the silage. It is also important to minimize the exposure of silage to air during storage and unloading.

Good quality silage has a pleasant odour, a typical colour and texture and a high nutritional value. Silage should be prepared carefully in order to assure successful ensiling fermentation. Plant composition should be adequate for that purpose. This includes adequate dry matter and water soluble carbohydrate content and low buffering capacity (resistance of the plants to pH decrease). High buffering capacity evolves from high soil, mineral and protein contents. If the crop is too moist or lacks enough fermentable sugars, or its buffering capacity is too high, secondary fermentation by clostridia will take place, resulting in increased pH (through the conversion of lactic into weaker butyric acid) spoilage and losses. Very moist crops (less than 30% dry matter) result in effluent which leads to loss of nutrients and to environmental pollution. Therefore, if possible, moist plants are wilted in the field to increase their dry matter content. If the crop is too dry fermentation will be too slow, consolidation might be incomplete and spoilage might occur by yeasts and molds.

Aerobic deterioration occurs when the silage is exposed to air which re-activates aerobic spoilage microorganisms such as yeasts and moulds. Some silages are more susceptible to aerobic spoilage than others. Aerobic deterioration is associated with loss of nutrients, decreased intake by animals, and sometimes, production of mycotoxins in the silage.

5.2.2 Methods of siling

The principle in making silage is to keep the green fodder material tightly packed in impervious containers, excluding air as much as possible. They should be harvested at the right stage of growth. If the crop is too succulent, it is preferable to allow it to get slightly wilted to reduce the moisture content before ensiling it, as this makes better silage.

The material is chopped up and packed into the "silos" with suitable machinery.

The container in which the silage is made is of greater importance and will determine to a larger extent the nature and quality of the final product. The size of the container will generally depend upon the number and kind of animal to feed from it and its height on the length of the feeding period. Different types of silo have been designed. Some of them are described below. The silo should be as near as possible to the crop to be ensiled. In the big dairy farms crop should be ensiled as near as possible to the stock. The land for making the silo should be at a high elevated place so that water does not stand and does not seep in.

Stacks

This is the simplest method of making silage, and it is really not a container at all. The stack should preferably be circular as this reduces wastage and the diameter should be adjusted to the size of the crop so that the final settled height is less than 2 m, the minimum diameter should be 3 m and there after settling will be 2 m high and will hold 25 tonnes of silage. In making the stack silage, the lowest layer should be allowed to warm up to 38^0C. This is especially necessary for young herbage. After first layer is warmed up the filling should be done rapidly. During filling up, trampling should be done in the middle. When the last layer has been added, the surface of the stack should be topped off with a shallow layer of old stacks, straw or rubbish. After that, the excavated soil should be put on the top of the curves and added upto a depth of 30 cm and weight should be kept to keep it compact as far as possible.

Advantages of stacks

- The stack is very economical to make. Can be built at the most convenient places, either in the field or where the silage has finally to be fed.

- It is best applied when immediate action is necessary. For instance, when the crop is likely to be ruined by continued bad weather or when the crop has been badly laid down by the storms.
- It needs no special equipment and can be done at a short notice
- When the silage has been used, the ground is left in its original state.

Disadvantages of stacks

- There is free access of air all around which is reflected in high temperature.
- Spoilage is highest when compared with other type of silage
- The crop cannot be trampled properly.

Clamp silo

In clamp silo, greater parts of crop remain above ground and the rest remains in slight excavated trench or pit. In addition, the clamp will be long and narrow, and low in relation to length. It is a compromise between the stack and the pit trench. In making a clamp silo, a shallow excavation is dug maximum up to 60 cm. The size of the clamp is maintained at 4 m of width and to a final height of 2 m. Each 30 cm of silo of the above mentioned dimension contains 1.5 tonnes of silage. If necessary, preservatives like molasses can be added.

Pit silo

The pit silo is cylindrical or rectangular and deep. Its shape is like that of clamp silo, but extends below the ground. The pit can be excavated from any suitable soil, not subjected to water logging. If silage is to be made annually, it is better to have a concrete floor, making provision for effluent to escape. A pit of average width of 4 m and with the silage settled to a depth of 2 m will hold 1.5 tonnes of silage for each 30 cm length. When filling is completed, an airtight seal can help to eliminate waste. Sealing can be done through polythene sheets. Weight can be applied by putting 10-15 cm of soil and this will provide necessary compression to the top most layer of silage. The loss of silage in this method is about 27 per cent.

Advantages of pit silo

- A pit silo is economical to build when the wall can be plastered
- Owing to its depth and shape, the pit silo has a large capacity for its size
- Less power is required for filling
- The smooth plastered wall allows the silage to settle uniformly and retain the juices
- Pit silo cannot be considered as permanent silos but if kept in good repair they will last indefinitely.

Disadvantage of pit silo

- It is not convenient to take out the feed
- The pit silo occupies farm land which is permanently inaccessible for cultivation, is tedious to build and require lot of labour
- The main difficulty is to ensure adequate compression, since, the depth of the silage is always out of proportion to the area covered, in strong contrast to the tower silo.

Trench silo

The difference between the pit silo and the trench silo is merely of size, the later usually having greater length in relation to breadth. The process of ensiling is more or less similar to that of pit silo.

Advantages of trench silo

- Cows and tractors are used to pack the silage
- Power required for filling the trench silo is less
- It is well adapted to ensilage of immature corn, cane and used emergency feeds
- The major part of the material converge will settle into the trench below ground level, the chance of air getting in at the edge is reduced to a minimum

Tower silo

It is round, cylindrical and is placed above the ground level. The height varies from 6 to 10 m or more with a varying diameter (6 to 10m). The erection of the silo is expensive. The materials used include wood, reinforced concrete or sheet metal. Use of wood is of much advantage in that silage acids do not affect it, on the other hand the wood tends to preserve it. For filling the silo a chopper blower is necessary. In this type, three type of silage are found (a) in the bottom it will be over compressed, sour and will give out smell of butyric acid (b) in the center it will be good, not too tightly packed and yet compressed well enough to give well preserved material and (c) in the top third, it is often dark and over-heated, near the surface and it will be of low value.

Advantages of tower silo

- Material can be preserved, with no side waste due to air leakage
- Wilting of crops and the sealing of silo are not as important as in pit silo, because the mass itself applies pressure and acts as an air seal to the lower layer
- The loss of dry matter is only 10 per cent

Disadvantages of tower silo

- It is very expensive to make
- Chopper blower is needed for filling up of the silo
- Emptying is very laborious
- In dry hot places the silage gets dehydrated

Temporary and portable silo

There is need for a small and easily portable silo where surplus grass can be ensiled and the produce can be fed outside. In such a silo upto 30 tonnes of silage can be preserved. A sectional polygonal silo made of wood and blotted together is light, efficient and can easily be moved. Of late, plastic PVC or polythene sheets and vinyl are used to enclose the silage completely. Vacuum can also be created by pumping such structures.

5.2.3 Silage additives

Silage additives can be used in order to enhance silage fermentation and their nutritional quality. They are classified according to their function, fermentation stimulants, fermentation inhibitors, aerobic deterioration inhibitors, nutrients and absorbents (McDonald *et al.*, 1991). Additives should be used according to needs and silage properties. It should be emphasized that additives can improve silage quality and minimize losses, but cannot compensate for poor silage making and management. Additives which come in a variety of forms like liquid, powders or suspensions. Additives can be applied during the harvesting, chopping operation or during filling of the silo.

Bacterial inoculants are used in order to enhance the ensiling fermentation. They are safe, easy to use, non-corrosive to machinery and regarded as natural products. Most commercial inoculants for silage include homofermentative lactic acid bacteria (e.g. *Lactobacillus plantarum, Enterococcus faecium and Pediococcus* spp.). They are used because they are fast producers of lactic acid. However, in whole crop cereal silages such inoculants resulted in spoilage upon aerobic exposure because their fermentation did not produce enough volatile fatty acids such as acetic acid, that inhibit fungi. The new types of inoculants are tailored according to silage properties, and for whole crop cereal silage heterofermentative inoculant is included (Weinberg and Muck, 1996).

Chemical additives include various organic and mineral acids that lower the pH artificially and inhibit specific microbial populations. For very moist crops formaldehyde can be used at 2-4 g l^{-1}. Other chemicals are available as well (e.g. sulphur based chemicals that inhibit fungi).

6. PROBLEM SOIL MANAGEMENT

6.1 Management of problem soils for forage cultivation

Possibility of extension of fodder area in the country is more remote because of preferential food need for human being. Fodder shortage of more than 40 per cent is often projected which calls for a concrete efforts in this regard. The country has presently about 6.9 million hectare of land under cultivated fodder crops and about 10 million hectare of land additionally would be required for meeting the fodder shortages. However, it is possible to capitalize the area available in the country in the form of problem soils which accounts for 25 million hectares as the soils can often may not be useful for other arable crops. Forage cultivation will be the best for economic exploitation of these soils.

The important problem soils are saline (5.50 million hectare), alkali (2.50 million hectare), water logged (6.00 million hectare), ravines and gullies (3.97 million hectare), shifting cultivation (4.36 million hectare), Riverine and torrents (2.73 million hectare). In addition to these, a sizeable degraded barren land in the mined area is also available in the country. These soils can be managed in two ways, one by suitable management of these soils for growing crops and secondly, by growing such plants which can make the best use of environment. Suitable forage crops and their varieties have been identified for many of such situations and given in Table.

Table 11. Crops and varieties suitable for problem soils

Nature of soil problem	**Crops**	**Varieties**
Acid soil	Cowpea (*Vigna ungigulata*) Rice bean Oats (*Avena sativa*) Deenanath grass (*Pennisetum pedicellatum*) Maize (*Zea mays*) Setaria (*Setaria sphacelata*) Napier-pearl millet hybrid Guinea grass (*Panicum maxicum*)	EC 4216, PUC 5286, UPC 4200 K1, K 16 Kent, KHO 810 Bundle Deenanath 1 Bundel Deenanth 2 African tall, PLP local A-de-Cuba Sel 125, Narok, PSS 1, Nandi IGFRI 7, IGFRI 10 Hamil, PGG 9, PGG 14
Calcareous soil	Sorghum (*Sorghum bicolor*) Pearl millet (*Pennisetum glaucum*) Napier-pearl millet hybrid Teosinte (*Zea mexicana*) Lablab beans	PC 6, HC 136 L 72, Rajko IGFRI 6, IGFRI 7 Improved, TL 16 JLP 4
Saline soil	Sudan grass Sorghum Oats Teosinte Napier-pearl milllet hybrid Lucerne (*Medicago sativa*) Senji (*Melilotus indicus*) Hedge Lucerne Shevri Para grass (*Bracharia mutica*) Rhodes grass (*Chloris gayana*)	SSG 59-3 HC 136, S 171 Kent, OS 6 Improved, TL 16 IFGRI 6, IFGRI 10 T 9 FOS 1
Coastal saline	Signal/Congo signal Coix sp Guinea grass Para grass Oats Barley (*Hordeum vulgare*)	 Hamil, PGG 14 Kent, JHO 851, JHO 822 DL 36, Azad

Contd...

Table 11. Contd...

Sodic soil	Berseem (*Trifolium alexandrinum*) Sorghum Oats Senji Napier-pearlmillet hybrid Karnal grass (*Leptochola fusca*) Hedge Lucerne Blue panic (*Panicum antitotale*) Para grass Rhodes grass	Mescavi PC 6, PC 23 Kent, JHO 822 Fos 1, Safed 76 IGFRI 6
Marshy/wet land	Teosinte Sorghum Para grass Congo signal Coix sp *Iseliema laxum* *Rhodes grass*	Improved, TL 1 PC 9
Degraded forests and ravines	Deenanath grass Guinea grass Stylo (*Stylosanthes hamata*) *S.guianensis* *Chrysopogon fulvus* *Subabul* *Prosopis juliflora* *Acacia nilotica*	Bundel Deenanath 1 Bundel Deenanath 2 Hamil, PGG 1, PGG 14, PHH 19
Desert/arid soil	Anjan grass (*Cenchrus ciliaris*) Yellow anjan grass Blue anjan Sewan grass	CAZRI 75, CAZRI 358 CAZRI 76 *FS 391* CAZRI 30-5

Source: Tripathi & Hazra, 1996

7. FORAGE SEED PRODUCTION

The seed production of forage crops is a specialized job requiring skill, intimate knowledge of the flowering and seeding behavior of the species, cutting and grazing management systems and the timeliness of the operations. All the tropical and subtropical forage species are by nature, shy seed-producers.

Table 12. Nutritive value of forage crops and feeds (% dry matter basis)

S. No.	Roughages	CP	DCP	TDN	Carbohydrates	Ether extract	Ash
	Straws						
1	Paddy straw	3.4	0.0	45.9	77.0	1.2	18.4
2	Wheat straw	3.0	0.2	48.3	83.8	1.0	10.0
3	Finger millet	6.5	0.2	50.0	78.7	2.1	12.4
	straw					1.5	
4	Bajra karbi	6.9	0.8	48.0	83.7	1.4	8.5
5	Sorghum karbi	4.9	.0	51.0	75.2	0.5	9.8
6	Maize straw	3.6	-	-			10.5
	Hays						
8	Moth hay	17.1	-	-	73.8	1.7	13.8
9	Guar hay	21.2	-	-	57.4	0.9	10.5
	Cakes						
10	Cotton seed cake	27.5	-	-	61.5	9.4	6.4
11	Groundnut cake	50.0	-	-	34.3	8.1	7.6
12	Sesame cake	42.70	-	-	32.8	6.5	12.6
	Meal						
13	Guar meal	42.9	-	-	46.2	5.2	8.8

7.1 Problems in perennial forage grasses seed production

i) Non-synchronized flowering

Forage grasses have the tendency of non-synchronized flowering within the population and within a panicle. Seeds would reach maturity in some spikelets while some of the spikelets would be in anthesis. Under such situation, fixing the optimum stage of harvest becomes difficult. Defoliation, adjusting irrigation and fertilizer application are essential to achieve synchronized flowering. Non-synchronized flowering has got an adaptive advantage. If unfavorable weather or stress occurs, only a portion of the reproductive unit will be affected and others will help in perpetuation.

ii) Poor seed setting

Forage grasses also suffer from the problem of poor seed setting. Weather condition during flowering has a significant role to play. Dull, cool and rainy period if coincides with flowering, poor seed set will occur. In forage legumes, less bee activity may cause poor seed set, owing to lack of pollination. Internal plant regulations sometimes lead to ovular abortion in legumes.

iii) Seed shedding

Most of the forage crops are perennial and flowering over a long period. The flower which opens first will mature earlier than others and the ripened seeds are not retained on the panicle. In guinea grass, complete shattering occurs 30 days after ear emergence. In the rachilla, which holds the flower, abscission layer is formed affecting the water transportation and hence the glumes get dried up and the seeds shed. To overcome this problem seed retention can be incorporated through breeding. Spraying of adhesives or glues to prevent shedding has been attempted on *Chloris gayana* and *Setaria* in Australia.

iv) Harvesting

Shattering nature of seeds poses problem in harvesting. Loss of seeds during harvesting operation is heavy. Non-uniform maturity results in incomplete recovery of seeds during threshing. Immature seeds possess high moisture and are susceptible to crushing and over dried seeds sustain cracking injuries during threshing.

v) Seed processing

Uneven maturity again poses difficulty in processing of seeds. Drying will not be uniform. Seeds of various maturity exhibit difference in size and so grading is difficult and may cause substantial rejection loss. Seeds also contain attached glumes and are with chaffy materials.

vi) Seed certification standards

Certification standards for some of the forage grasses require revision since, they are unachievable especially the seed purity standards.

vii) Seed dormancy

Dormancy in forage seeds is caused by tight glumes and inhibitors present either in glumes or in seeds. Forage legumes invariably posses hard seed necessitating scarification treatments before sowing.

7.2 Agronomic requirements for high seed yields

Like other cash crops, the forage crops show a good response to higher levels of nutrient supply for a higher out-turn of seed both in the case of cultivated and pasture species.

(a) Effect of nitrogenous fertilizers

All the tropical and subtropical grasses show a linear response for seed production to the application of nitrogen. In *Cenchrus ciliaris*, application of 670 kg of N per ha increased the seed yield from 8 to 500 kg per ha in Australia. Similarly, seed yield of the order of 20-60 kg per ha were obtained in the case of para grassa (*Brachiara mutica*) with 240 kg of N ha^{-1}. In Golden Timothy (*Setaria sphacelata*), 70 to 120 kg ha^{-1} of seed was obtained with 330 kg of N per ha. Likewise in India, the application of 40 kg of N per ha in two split applications during monsoon increased the seed yield from 17 to 30 kg per ha in the case of *C. ciliaris*, 23 to 143 kg per ha in the case of *C. setigerus* and 70 to 239 kg ha^{-1} in the case of moshi (*Iseilema laxum*).

(b) Effect of phosphatic fertilizers

Phosphatic fertilizers have a greater tendency to affect the fodder and seed yield, especially in the case of leguminous crops. Foliar spraying with 4 per cent phosphoric acid (40 kg of P_2O_5 ha^{-1}) in the case of berseem, 15 days after the third cut in February resulted in 50 per cent increase in seed yield over that from the control. Similarly, in cowpea (*Vigna sinensis*) 50 kg of P_2O_5 ha^{-1} as soil application at the time of sowing in July coupled with three split foliar sprays of 2 per cent phosphoric acid (25 kg of P_2O_5 ha^{-1}) starting from the first flush of flowering and at an interval of 3-5 days, increased the seed yield by 30 per cent.

(c) Rational use of irrigation

Seed-setting in the case of cash crops and forage crops is influenced by the available soil moisture at the critical periods of plant growth, i.e., flowering and anthesis. Taking advantage of this fact, water can be economized greatly and made available for other crops. Thus in case of berseem, application of two irrigations at 30 and 45 days after the third cutting resulted in a slightly higher seed yield than with four irrigations given at the recommended 15 days interval, starting from the third cut. Likewise, the seed production in Kent oats, with two irrigations (after the crop establishment at 20 and 95 days growth) was of the same order as with five irrigations applied at 20, 45, 70, 95 and 115 days of growth.

7.3 Physiological approach to seed yield

The forage crops because of their excessive vegetative growth fail to produce sufficient seed for further multiplication on an appreciable scale. Some of the growth regulators play an important role in increasing the seed yields. Foliar spraying with Cycocel (CCC), a growth-retardant, found to enhance seed production in case of oats and jowar (M.P. chari). Spraying with CCC @ 8.5 kg a.i ha^{-1} at about 45 days growth in the case of M.P. chari and 2.0 kg a.i ha^{-1} just before the elongation stage in the case of Kent oats increased the seed yield from 6.3 to 15.4 q ha^{-1} in the case of M.P. chari and from 1.8 to 27 q ha^{-1} in 'Kent' oats.

Similarly, a composite spray of micronutrients (Cu, Zn, Mo, B and Mn) in the form of zinc sulphate, manganese sulphate, and iron sulphate (each 100 ppm) and borax and sodium molybdate (each 10 ppm) at the rate of 700 litres per ha resulted in 30-50 per cent increase in seed yield in berseem.

Flower abscission in case of crops such as lucerne, berseem, cowpea, velvet bean, field bean, etc. is very common, but the growth regulators when applied in low concentrations have the tendency to check abscission and produce more flower bud primordia, resulting in good seed set per fruiting pod and a higher number of pods per plant. Thus the spraying of B-nine and CCC (each at 500 ppm) and Phosfon (100 ppm) to cowpea, lucerne and berseem increased seed yield by 60-90 per cent.

7.4 Role of external agencies

Other extraneous factors, such as the optimum population of efficient insect-pollinators in case of cross-pollinated crops, such as lucerne and berseem, play a very significant role in seed production. It has been observed that temperate clovers, i.e. white clovers, can be pollinated by all kinds of bees, but bumblebees work much faster in red clover than the hardy bees. The efficiency in pollinating the flowers of lucerne also varies. It has been observed that the honeybees do not trip the flowers as efficiently as some of the other insect-pollinators, such as *Ceratina binghami* and *Megachile flavipes*.

In the case of legumes with a trailing growth habit, *viz.*, cowpea, velvet bean, field bean, *Siratro, Clitoria ternatea, Atylosia*, etc., the seed setting and pod development is poor because of the competition for light, since the pods after development get buried in the vegetative mass. Thus, staking of the plants or making them climb on the fences and other suitable structures increases the pod and seed set two to three times in field bean, velvet bean, siratro and cowpea.

7.5 Seed treatment

1. Forage grasses

a) Guinea grass

Freshly harvested seeds germinate only up to a level of 5 per cent, increase to 24 per cent after one year storage. If fresh seeds are treated for 10 minutes in conc.H_2SO_4, germination can be increased to 40 per cent.

Germination improvement is also possible by treating the seeds with KNO_3 solution.

By leaching the seeds in running water, germination can be improved

Seed treatment with GA 500 ppm or 0.1 per cent, Thiourea or KNO_3 increased germination.

b) Cenchrus sp

Seeds are soaked in water for 48 hrs to remove inhibitors

To overcome dormancy, the seeds are treated with GA_3 250-750 ppm, KNO_3 0.75 per cent or Thiourea 0.75 per cent.

c) Deenanath

Seed dormancy prevailed up to 16 weeks. Seeds from late formed tillers had longer period of dormancy. Soaking the seeds in 1:1 mixture of KNO_3 (0.25 per cent) and GA_3 (200 ppm) effectively overcome dormancy and gave 65 per cent germination.

2. Forage legume

a) Desmodium

Hard seed occur up to 22 per cent. For this acid scarification for 2-3 minutes is adopted which improves germination from 69 to 99 per cent.

b) Desmanthus

Acid scarification with conc.H_2SO_4 50 ml kg^{-1} of seed for 15 minutes

Hot water soaking at 70 ^{0}C conditions for 5 minutes

Aberration of seed for 15 minutes

c) Stylosanthes

Scarification with conc. H_2SO_4 @ 200 ml/ kg of seeds for 4 minutes

Soaking the seeds in GA_3 50 ppm for 3 hrs

d) Leuceana leucocephola

Scarification with conc. H_2SO_4 @ 100 ml kg^{-1} of seeds for 20 minutes.

e) Acacia nilotica

Physical scarification with sand paper

Soaking seeds in cold water for 48 hrs

Soaking seeds in hot water for (80^{0}C) for 30 minutes

Soaking seeds in cow dung solution

Acid scarification with conc. H_2SO_4 for 10 to 30 minutes.

Inseparable weed seeds

Lucerne seeds often get mixed with cuscuta seeds. They are similar in appearance causing problem in separation. In western countries, specialized equipment called Dodder mill is employed for separating lucerne and cuscuta seeds. Sieving of the seeds of lucerne with BSS 14X14 square hole sieves, makes the cuscuta seeds to fall through the sieve by virtue of its small size.

Seed pelleting

Seeds of forage crops can be pelleted with DAP and micronutrients like $FeSo_4$ or $ZnSO_4$ at 100 mg kg^{-1} of seed. Starch 5 per cent can act as adhesive and gypsum or arappu leaf powder can be the filler material. Sowing of pelleted seeds results in improved seedling vigour and enhanced crop performance.

Seed hardening

Among the chemicals tried, K_2HPO_4 1% solution was found to have influence on germination and growth of grass. Direct seeding of grass seeds behind the plough was the best. Hardening of leguminous fodder seeds, followed by 1 hour soaking in 1/3 volume of solution with 1% $FeSO_4$ maximized the yield of quality seeds.

Hardening with 1% *Prosopis* leaf extract was found to be beneficial and it was on par with $FeSO_4$ hardening.

Hardening followed with halogenation was superior to simple hardening.

Seed certification standards under Indian Seed Act (1996), for seed certification have been fixed for various forage grasses and legumes by the central seed committee.

Questions

1. List the forage groups with examples
2. Give the classification of forage crops with examples
3. Discuss in detail any two forage grass
4. Give the agro techniques for forage grasses
5. Explain the importance of legume forage and give the package of practices for lucerne and cowpea
6. Elaborate the importance of tree fodder in animal nutrition
7. Write in detail about the problems encountered in the perennial forage grass production
8. Enlist agro techniques for higher forage seed yield
9. Give seed treatment and seed hardening techniques for forage crops
10. Give short note on preservation of forage crops
11. Enumerate the factors governing quality of forage

—x—

Chapter 3

Farming System

1. IMPORTANCE

The population in most tropical countries is increasing leaps and bounds, warranting more production of food from limited cultivated lands. Many challenges are faced in our quest to achieve sustainable food security with shrinking land resources for agriculture to produce additional food grains to meet the requirement of the prognosticated population in our country. Furthermore, as agriculture becomes more intensive, soil and waste degradation threaten the sustainable agricultural production. India would need to produce additional food grains of 100 and 160 million tonnes by 2030 and 2050 AD, respectively, to feed its projected population.

India's population as on date is about 1003 million and is expected to reach 1370 and 1600 million in 2030 and 2050 AD, respectively. Food grain production in In1dia reached an all time high of 218 million tonnes during 2000-2001 and 289 and 349 million tonnes are needed to meet the demand of the projected population in 2030 and 2050 AD, respectively. The current net cultivable area of

142.8 million ha is likely to dwindle further by virtue of diversion of some of the cultivable area to domestic and industrial purposes. It is anticipated that the land area available for cultivation in 2050 would be 137 million ha.

Thus, our production of food grains per unit area has to almost double from what we are obtaining today. This could be made possible by putting the land, both irrigated and rainfed under intensive cultivation. Fortunately, most of our states lie in tropics and so are blessed with abundant solar energy thus making cropping possible round the year. The only way to increase agricultural production is to increase the productivity per unit area per unit time. In the scientific era of agriculture, cropping systems, genotypes, geometry of planting and management practices are designed to increase the productivity per unit area per unit time, simultaneously making efficient use of available resources and stabilizing yields.

More than 85 million, out of 105 million operational holdings in India are with less than two hectares of farm size, necessitating intensive cultivation involving the modern concept of time and space by adopting heavy dose of chemical fertilizers leading to accumulation of chemical residues in the soil (Swaminathan, 1987).

Compounding the problems arising from the growing pressure of human population on the carrying capacity of the land and productivity, the agricultural systems have met with various kinds of ecological and environmental problems. The ultimate goal of sustainable agriculture is to conserve the natural resource base, protect the environment, and enhance the health and safety of human population over a longer period. This can be achieved by seeking the optimal use of internal production inputs in a way that provide acceptable levels of sustainable crop productivity and livestock production resulting in economically profitable return.

The average holding of a farm in India has been declining and as indicated earlier over 80% of operational holdings are below the size of 1.00 hectare (Mahapatra and Bapat, 1992). There is no scope for increasing the farm size, because of steady increase in population with shrinkage of cultivated land as a result of industrialization and urbanization. Only vertical expansion is possible by integrating appropriate farming components requiring lesser space and time ensuring periodic income to the farmer. The integrated farming systems, therefore, assumes greater importance for sound management of farm resources to enhance the farm productivity, reduce the environmental degradation, improve the quality of life of resource poor farmers and to maintain the sustainability. Hence, the sustainable farming systems, economically viable and ecologically compatible encompassed with higher productivity to meet the present and future needs without jeopardizing the potential, are to be optimized for specific agricultural domain.

Integrated farming system is a judicious mix of one or more enterprises with cropping in which there is a complementary effect through effective recycling of wastes / residues and encompasses additional source of income to the farmers.

It would be possible to reach the high level of productivity, be inherently more sustainable with proportionately less input through integrated farming systems. This helps in utilizing the waste of one component as input of another leaving behind nutrient rich organic end products for sustaining the soil productivity. Therefore, the choice of components in the farming system and their management must ensure that the interaction between components should be the most complimentary with least competitiveness. Even though the progress of Indian economy is very much linked to the advancement of agriculture, different farming systems have been developed independently and are being practiced by farmers indigenously without any rationale for utilizing the waste/ residue arising out of cropping and other related activities. This leads to the economic imbalance by virtue of improper tapping of the potential of each component in the system and creating an impression among the farmers that the system is not viable between and within the farm holdings.

Integrated farming system approach is not only a reliable way of obtaining fairly high productivity with considerable scope for resource recycling, but also a concept of ecological soundness leading to sustainable agriculture. With increasing energy crisis due to shrinking of non-renewable fossil-fuel based sources, the fertilizer nutrient cost have increased steeply and with gradual withdrawal of fertilizer subsidy, it is expected to have further hike in the cost of fertilizers. This will leave the farmers with no option but to fully explore the potential alternate sources of plant nutrients atleast for the partial substitution of the fertilizer nutrients for individual crops and in the cropping systems. Integrated nutrient supply approach for the cropping systems by judicious mix of bio resource residues along with fertilizers has number of agronomic and environmental efficiencies. The success of integrated agricultural allied enterprises with cropping for different farming situations has been studied by many researchers under different agro-eco systems.

2. FARMING SYSTEM

With a shift from crop to cropping systems approach in agricultural research, development and extensive usage of large number of new terms have been phrased and defined by various workers. However, many of them call for precise and universally usable definitions in all the types of agro climatic and edaphic situations.

A farming system is a collection of distinct functional units such as crop, livestock, processing, investments and marketing activities which interact because of the joint use of inputs they receive from the environment which have the common objective of satisfying the farmers' (decision makers) aims. The definition of the borders of the options depends on circumstances; often it includes not only the farm (economic enterprise) but also the household (farm - household system)".

Ruthenberg (1971)

Farming system is as the way in which farm resources are allocated subject to the needs and priorities of the farmer in his local circumstances. The farming system is more risky than any other systems and specifically refers to a crop combination or enterprise mix in which the products and byproducts of one enterprise serve as input for the production of other enterprises.

Collinson (1979)

A farming system is not simply a collection of crops and animals to which one can apply his input or expect immediate result. Rather, it is a complicated interwoven mesh of soils, plants, animals, implements, workers, other inputs and environmental influences with the strands without link help and manipulated by a person called farmer, who given his preferences and aspirations, attempts to produce output from the input with the technology available to him. It is the farmer's unique understanding of his immediate environment, both national and socio economic that results in a "Farming system". Thus, Farming system (farm system or whole farm system) is the production and consumption activities used by a person called a farmer to derive benefits from land and other inputs through crop growth and the use of technologies available to him under specific environmental conditions.

Zandstra *et al.* (1981)

Characterizing farming systems is exceedingly difficult due to the complexity and heterogeneity of the components involved. Farming system methods are compatible with traditional research approaches and have evolved as a means for involving farmers and farm families in setting research priorities and in identifying appropriate paths to agricultural development. Specifically, farming system views the whole farm as a system and focuses on interdependencies, between the system components and their interaction with physical, biological and socioeconomic factors where more commonality would be found within the system than between systems. Therefore, farming system is part of larger systems and that would be divided into many sub-sub systems.

Shaner *et al.* (1982)

Farming is the process of harnessing solar energy in the form of economic plant and animal products, and 'System' implies a set of inter related practices/processes organized into a functional entity, *i.e.* an arrangement of components or parts that interact according to some process and transforms inputs into outputs. Faming system is a decision making unit comprising farm household, cropping and livestock systems that transform land, capital and labour into products for consumption and sale.

Fresco and Westphal (1988)

Farming system is a set of agro-economic activities that are inter-related and interact among themselves in particular agrarian settings. Farming system is the mix of farm enterprises to which farm families allocate its resources in order to efficiently utilize the existing enterprises for increasing the productivity and profitability of the farm. These farm enterprises are crop, livestock, aquaculture, agro-forestry and agri-horticulture.

Sharma, *et al.* (1991)

The interaction of biological and economic factors together shape farming systems. Farming system is simply an orderly set of interdependent and interactive components, none of which can be modified without causing a relative change elsewhere in the system.

Chinnusamy and Purushothaman (1992)

Farming system is a mix of farm enterprises such as crop, livestock, aquaculture, agro-forestry and fruit crops to which farm family allocates its resources in order to efficiently manage the existing environment for the attainment of family goal in other words.

Pandey *et al.* (1992)

Farming system is an integrated resource management strategy for obtaining economic and sustained crop and livestock production and preserving the resource bases with high environmental quality.

Palaniappan (1992)

Farms are complex ecosystems involving interactions between the components and processes of natural systems and resource allocation management and conservation practices associated with crop and livestock production.

Steve Oberle (1994)

These definitions though comprehensive, leave one or the other aspect untouched, call for rephrasing for extensive communicability, usage and applicability and hence the following is a comprehensive definition.

"Farming system represents an appropriate combination of farm enterprises *viz.*, cropping systems, horticulture, livestock, fishery, forestry, poultry and the means available to the farmer to raise them for profitability. It interacts adequately with environment without dislocating the ecological and socio-economic balance on one hand and attempts to meet the national goals on the other. The farming system in its real sense will help in different ways to lift the economy of agriculture and standard of living of the farmers of the country as a whole.

2.1 Farming Systems Research

Farming Systems Research method is designed to understand farmer's priorities, strategies and resource allocation decisions. It is most often used in conjunction with On-Farm Research to identify and adopt technologies useful to location specific problems of farmers.

Considerable attention is currently being given for improving small farmers in developing countries. An important way of helping them is through agricultural research, extension and related programmes specific to their needs. Research and development programmes had often been undertaken without having small farmers in mind or without knowing much about them. In contrast, the Farming Systems Research and Development approach starts and ends with small farmers and thereby focuses specifically their conditions and aspirations.

The goal of farming system research is to develop sustainable land use systems which will optimize the farm resource use, minimize degradation with consideration to regenerative capacity and increase income and employment for farm families and promote quality of life and environment.

2.2 Concept and principles of Farming System Research

The modern agriculture emphasizes two more dimension *viz.*, time and space concept. Time concept relates to increasing crop intensification in a situation where there is no constraint for inputs including irrigation. In other words increasing the cropping intensity in areas where the production potential *viz.,* land is underutilized even with full resource potential. It is a time bound programme, where for most of the field crops it is considered for 365 days or 12 months. In case of long duration crops and perennial crops, the duration of each rotation will vary from two to three years depending upon the duration of constituent crops. The areas where, only one crop (100%), two crops (200%) and three crops (300%) are raised in a year, leaving the land fallow for two to eight months, the cropping intensity has to be increased to 200%, 300% and 400%, respectively, well within the same one year. This will provide an opportunity to increase gross cultivable area in a year and will ultimately help to enhance the productivity per unit area per unit time.

In case of rainfed areas where there is no possibility of increasing the intensity of cropping, the other modern concept *viz.,* 'space concept' can be applied. In space concept, crops are arranged in a tier system combining two or more crops with varying field duration as inter crops by suitably modifying the planting method. By way of building up of crops/other agricultural allied components in a vertical dimension (otherwise called multi-tier system) the land equivalent area can be increased. Thus by making use of these time and space concept either in irrigated or in rainfed areas within a specified time span say a year, and unit area of land, a hectare, productivity is sought to be increased by repeated and or intensified cropping.

This calls for urgent action on the part of Agronomists and Scientists of related disciplines to devote their attention on the design and testing of cropping systems for different regions. Income through arable farming alone is insufficient for bulk of the marginal farmers. Activities such as dairy, poultry, fish culture, sericulture, biogas production, edible mushroom cultivation, agro forestry, agri-horticulture etc., assumes critical importance in supplementing their farm income. It fits well with farm level infrastructure, and ensures fuller utilization of by products.

The farming systems research strategy emphasizes that the research agenda should be determined by explicitly defined farmers needs through an understanding of the existing farming systems rather than its perception by the researchers. It is a holistic approach with the main focus on strengthening the partnership between the researchers, extensionists and farmers. It provides an opportunity for the farming community to influence the research agenda to generate relevant technologies.

Realizing the need to tailor research to meet client demand, many programmes have been initiated in Agricultural Research in recent years with an orientation and focus on farming systems. The main endeavor has been to integrate the desirable features of the complex farming systems research approach into the mainstream agricultural research so that the technologies developed are relevant, client oriented and location specific, This demands initiative and innovativeness on the part of researchers and the system. It is later realized that Farming System Research is not the only means for all the ills of agriculture, but can certainly complement the conventional commodity oriented research strategies in generating and disseminating appropriate agricultural technologies.

Integrated Farming System seems to be the only answer to the problem of increasing food production, for increasing income and for improving nutrition of the small-scale farmers with limited resources. In a cropping system, the amount of byproducts can be as high as or higher than the marketable produce. This may go as waste if not utilized in an animal enterprise. Hence, integration of different agricultural allied enterprises with crop activity as base, will provide ways to recycle produces/waste materials of one component as input on the other linked component and to reduce the cost of production of the economic produce of component and finally to enhance the net income of the farm as whole.

2.3 Primary objectives of Farming System Research

- To identify the constraints in increasing farm productivity
- To provide technological intervention options for improving farming systems at a given resource base through farmer participatory research
- To conduct farmers participatory research in refining technologies of farming systems and provide feed back of farm problems to on-station researcher
- To monitor the impact of component or system related technology on sustainability of farming systems to meet the growing needs of population with emphasis on equity and gender issue

2.4 Farming System Research interrelated objectives

- To understand the physical and socio-economic environment within which agricultural production takes place
- To gain an understanding of the farmer in terms of his/her skills, constraints, preferences and aspirations
- To comprehend and evaluate existing important farming systems, in particular, the practice and performance of these systems.
- To improve the identification of problems and opportunities for change in existing farming systems and thereby better focus of research on specific key aspects

- To enhance the capacity of research organization to conduct research on priority problems
- To conduct research on new or improved practices or principles and to evaluate these for possible testing on farms
- To evaluate new or improved systems or system components, on farm in major production areas under normal farm conditions
- To assist the extension, monitor the adoption, and assess the benefit of improved farming systems

These objectives can be met within the context of three multidisciplinary "activity areas" - the data base and database analysis, research station studies (On-Station Research), and On-Farm studies.

- Data-base analysis involves the collection, collation and analysis of data on many factors characterizing the environment and farming system of a region
- On-Station Researches involve a focused research programme aimed at the development of components for improvement of existing systems or for the putting together of new systems
- On-farm researches involve studies of existing farming systems, on farm experimentation, studies of technology adoption and assessment of the impact of new technology-all in relation to the farm household. Framework of farming systems research Fig. 1

2.5 Steps in Farming Systems Research

- *Farm selection:* Select the agro-ecological zone in which FSR is to be initiated. If necessary, further divide this agro ecological zone to identify specific farming situation

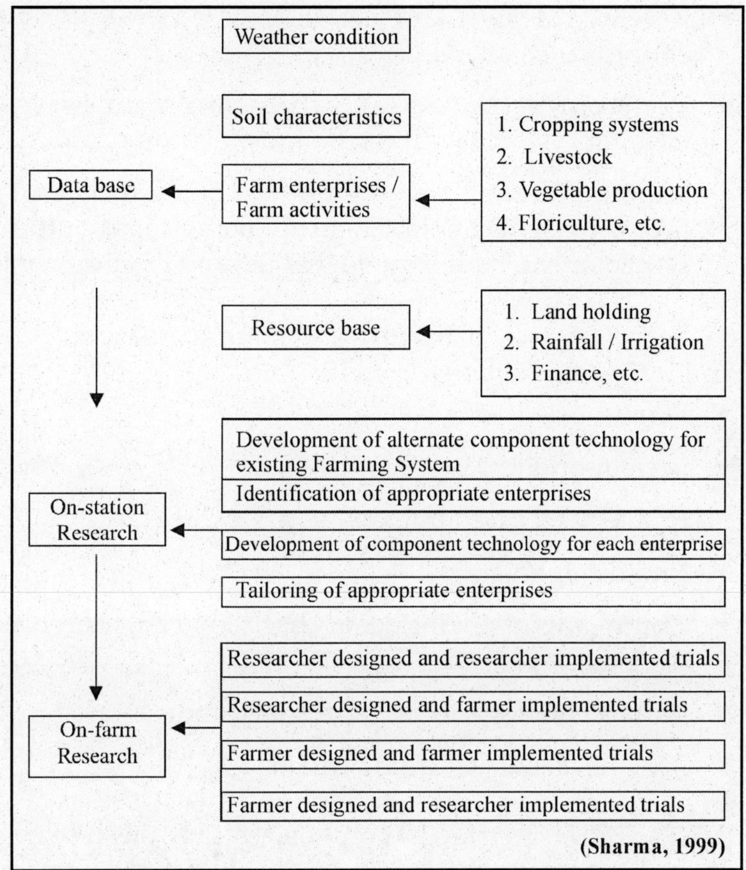

Fig. 1: Framework of farming systems research

- *Selection of villages and farmers:* Select villages in each farming situation comprising marginal/small and medium/large farmers. Selection of village and farmers should be at random so as to represent all farming community of the target area
- *Diagnosis of constraints in increasing farm productivity:* Carry out survey through rapid rural appraisal. Prepare an inventory of farm resources and support services. Identify the production constraints.

- Research design and technology generation and adoption.
- Technology transfer and diffusion of improved farming systems within recommended domain.
- Impact of Technology of Improved Farming System - productivity, Economic return, Energy input-output, Employment, Equity (gender issue) and Environment.

2.6 Characteristics of Farming System Research

Farming System Research activities are

- Farmer-oriented
- System-oriented
- Problem-solving approach
- Interdisciplinary
- Complements mainstream disciplinary research, it does not replace it.
- Tests technology in on-farm trials, and
- Provides feedback from farmers.

The FSR & E strategy emphasizes that the research agenda should be determined by explicitly defined farmers' needs through an understanding of the existing farming systems rather than its perception by the researchers. It is a holistic approach with the main focus on strengthening the partnership between the researchers, extensionists and farmers through extensive on-farm activities. It provides an opportunity for the farming community to influence the research agenda to generate relevant technologies

Realizing the need to tailor research to meet client demand, many programme in the Indian Agricultural Research System have been initiated in the recent years with an orientation and focus on farming systems. The endeavor has been to integrate the desirable features of the complex FSR & E approach into the mainstream agricultural research so that the technologies developed are relevant, client-oriented and location-specific. This demands initiative and innovativeness on the part of the researchers and the system. It is realized that FSR & E is not panacea for all the ills of agriculture, but can certainly complement the conventional commodity-oriented research strategies in generating and disseminating appropriate agricultural technologies.

FSR & E has been perceived variously and in different contexts, but it has certain distinctive core characters. According to Shaner *et al.* (1982), farming system would take into account the interactions between subsystems within a whole farm setting and is, thus, designed to address farmer and society needs and goals in their entity.

Core characters of Farming System Research and Extension are:

a) *It is holistic:* The whole farm is viewed as a system encompassing interacting subsystems. No potential enterprise is considered in isolation. It looks at the farm family household as a system of natural and human components.

b) *It is farmer participatory:* Involves farmer participation in different stages of technology generation and dissemination including systems description, problem diagnosis, search for appropriate technology, design of on-farm research, implementation of on-farm trials, monitoring and evaluation, feedback and diffusion.

c) *It is problem solving:* As an applied problem solving approach, it emphasizes research aimed at developing technologies to relax production constraints. The research process begins with diagnosis of problems and tracing of their biophysical causes and socio - economic and institutional constraints that influence technological solutions.

d) *It is gender sensitive:* Explicitly acknowledges gender-differentiated roles of farm family in agriculture. The user perspective in technology development is emphasized by reviewing farming systems in terms of activities analysis, access and control over resources (including knowledge) and benefits, and implications in developing relevant research agenda.

e) *It is interdisciplinary:* Adopts interdisciplinary approach to solve agricultural problems that are of concern to farmers. It lays greater emphasis on interdisciplinary cooperation among the biological and social scientists.

f) *It is interactive and iterative:* Rather than trying to know everything about a system, it requires knowledge of only a few key functional relationships to produce a useful analysis. Interactions with farmers at each step of technology generation are followed sequentially. But approach is intended to be iterative, a recurrent analysis of the farmers' system, permitting continuous learning and adaptations.

g) *It emphasizes extensive on-farm activities:* On-farm research with a farming systems perspective emphasizes search of multiple options for problem solving. The objective is to structure appropriate situations where farmers could express their preferences and apply their evaluation criteria for selecting technologies. Farmers, researchers and extensionists work as partners influencing each other.

h) *It complements experiment station research:* FSR & E is not being promoted as panacea for all maladies of local agricultural production systems. Identified technologies could be adapted to specific situations through adaptive trials or technology verification trials. This, in turn, may require further on-farm trials for technology refinements and adaptations to micro ecological niches.

i) *It acknowledges the location specificity of technical solutions:* This also implies that two kinds of issues be addressed through the FSR & E process, *viz.*, (a) the conditions under which the basic components of the technology can be made to work, and (b) conditions under which the systems technology interests different groups of technology users.

j) *It recognizes interdependencies among multiple clients:* Productive, interactive linkages among policy planners, researchers (on-station and on - farm), extension and development agencies and farmers are critically important in contributing to the process of agricultural development.

k) *It emphasizes feedback:* Great importance is placed on feed back to i) modify the content of subsequent on-farm trials; ii) initiate new experiments and even change research priorities; and iii) effect policy shifts based on microlevel analysis. It is this dynamic interplay between multiple constituencies that can help realize the potential of FSR & E.

(Raman and Balaguru, 1991)

2.7 Issues in Farming System Research and Development (FSR & D)

At a policy making level, a range of issues must be addressed in deciding whether to use an FSR & D approach. The basic issue concerns identifying which conditions are most conducive for organizing research and development along FSR & D lines.

Most policy makers are interested in increasing the returns from their investments in agricultural research. Therefore, the question arises as to whether satisfactory returns are being secured from current investments in research and development. This is particularly important in most developing countries, where shortages of scientific personnel and insufficient research budgets restrict agricultural research to only modest efforts. Within this context, policy makers must be continuously alert to ways in which research policies and programs can respond better to national needs.

A review of the conditions in a number of developing countries suggests that agricultural research and development are undergoing important changes as governments seek to enhance the income and productivity of small farmers. The character and origin of these changes can be brought out by asking questions about society's goals and small farmers' needs.

(Palaniappan, 1994)

a. Time required to obtaining results

The general approach to FSR & D is repaid initiation of on-farm experiments combined with adjustments in the programme direction as results provide feed back. With adequate planning, researchers often start experiments without missing a cropping season. Sometimes, they try explanatory experiments to learn how farmers respond to new opportunities: at other times, researchers conduct trials to screen locally available technologies for their applicability to specific farmer's conditions. Under favourable conditions, some research results may be ready for widespread diffusion to farmers with in a season. However, more fundamental changes in farmers cropping pattern and management practices normally take longer. To include, livestock as subsystem in integrated farming, animals such as cattle, goat, sheep, the environment, livestock systems and growth stages often require more careful study than most crops or small animals.

b. Organizational flexibility

FSR & D is primarily a modification of existing research and extension methods, therefore, the approach is adaptable to a variety of situations, as illustrated by the following possibilities.

- A country can implement FSR & D through a semiautonomous government corporative that has more flexibility in operation, budgeting personnel management, ministerial research and development.
- A country can implement FSR & D through Ministry of Agriculture if the Ministry is responsible for research and extension.
- A country can apply FSR & D to the activities of an experiment station in which one or more teams trained in FSR & D methods work closely with experiment station staff.
- A country can try FSR & D into project to increase production, in such a case; FSR & D methods can improve the efficiency of the overall project.

Each approach has its advantages and disadvantages. So the approach selected depends on the situation. Here, we emphasize that FSR & D, whether in whole or in part, can be and has been implemented in a variety of ways.

c. Staffing requirements

FSR & D strongly emphasizes working with farmers in their fields. To the extent that this emphasis is now, those currently at research stations or at regional or national headquarters will require some reorientation. This reorientation includes research methodology as applied to field conditions and methods for working with the whole farm family - male and female, young and old. Where appropriate, females may need to be added to the research and extension staff.

However, FSR & D does not replace existing research or extension; rather, it builds on the existing base;

consequently, experienced researchers and extension specialists usually remain in their existing organizations and much on the field staff consisting young professionals trained specifically for FSR & D's purpose. One approach is to begin FSR & D activities in one or two regions and after several years of experience, to choose leaders from these teams when moving to new areas. Heads of FSR & D programmes must also train staff to replace those who periodically leave the programme.

d. Training

An early activity when implementing an FSR & D approach is to train the staff about the objectives, processes and methodologies of FSR & D. Training materials will need to be collected from on - going programme elsewhere and augmented by new materials appropriate for the country. During this early stage, the International Agricultural Research Centres (IARC's) and organizations with similar activities can be especially helpful.

The Principle objectives of the training are;

- to acquaint team members with on-farm techniques.
- to give them guidance and experience for working as an interdisciplinary team.
- to instill in the team members an enlightened appreciation of small farmers as a useful source of information and as valuable partners in the research and implementation process.

Initially, program leaders may want to take advantage of production and farming systems training at one or more of the IARC's and any regional center specializing in applied agricultural research with such training as a base, in country training program for both research and extension personnel can then be developed and implemented. Some staff members may be selected and sent abroad for further academic training.

e. Costs

A discussion on the relative costs of FSR & D centres deals with expenditures, rates of adoption and breadth of coverage. The first of the three issues concerns expenditures or facilities and operating costs. To the extent that FSR & D reduces experiment station activity, costs of expanded installations, operations and the accompanying staff will be lowered. However the field work requires increased expenditures for vehicle purchase and maintenance of field equipment overall, the combined initial and recurring costs of FSR & D appear to be lesser than the costs of comparable levels of activities on experiment stations when administrators consider the costs of building, staffing and equipping the stations. However, such comparisons are of limited value.

The second issue concerns the generation of new technologies acceptable to farmers. This is not a straightforward issue because the target group for FSR & D is sometimes different from that of general agricultural research.

The third issue centers on the range of applicability of research results. Traditional research, by its nature, often has general and wide applicability. FSR & D is designed to be more specific, but it may also be applied broadly if the team can identify environmental conditions sufficiently wide ranging and target groups in sufficiently large numbers.

f. Governmental support

Because FSR & D concentrates on field activities, the government will need to take steps to allow team members to efficiently carryout this work. Materials for conducting experiments need to be available at appropriate times; otherwise the experiments may not be completed. Reliable transportation is essential, especially where the terrain and weather conditions make travel difficult. Finally, incentives are often required to attract and hold qualified staff, incentives such as the recognition of team accomplishments will be needed to overcome the uncertainties of working in a new and different program and the hardships of living and working in remote areas.

FSR & D does not place great demands on the government but these demands must be met to create and maintain the momentum necessary to sustain an effective FSR & D effort.

g. Policy issue

Policy issues are inevitably an integral part of FSR and should be recognized as such. The problems of agriculture in developing countries are too great and too urgent to restrict the search for solutions to the set of possible "technological fixes" when so often and so evidently other features of the system, broadly defined, must also be changed before real progress is attainable.

3. INTEGRATED FARMING SYSTEM

3.1 Specialized Farming System

Specialization involves the intensification of the agricultural activity aimed at maximization of the production/unit area/unit time. This involves improvement of operational efficiency and speed of operation/execution at each step. The specialized farming system is thus focused on single cropping system or sequence of farming enterprise like animal breeding, dairying so as to achieve the "Highest degree of Precision Management" with minimal diversion of resources/attention to diverse crops or enterprises, differing widely in managerial skills and involving extremely different implements and/or handling capabilities for the produce.

3.2 Integrated Farming System

Integrated Farming System (IFS), a component of Farming System Research (FSR), introduces a change in the farming techniques for maximum production in the cropping pattern and takes care of optimal utilization of resources. The farm wastes are better recycled for productive purposes in the integrated farming system.

Unlike the specialized farming system (SFS), Integrated farming systems activity is focused round a few selected, interdependent, interrelated and often inter-linking production systems based on a few crops, animals, and related subsidiary professions.

Integrated farming is not a new concept, crops, trees and animals have traditionally been raised together on small family farms throughout the world. The demise of this concept in developed countries in modern times had led to extinction of small family, as crops, trees and animals are now raised separately and managed on large holdings in these countries. So the interest and expertise in the integrated farming system decreased almost all over the world as practitioners, scientists and educators focused of their own specific commodity and commodity oriented research.

Though the modern farming technology is not suitable for developing countries as small farms are inevitable because of large population, these countries are following suit as yields become higher due to use of higher inputs. The availability of improved seeds and breeds and supply of high inputs like chemical fertilizers, irrigation and pesticides have led to perceptible increase in crop production, specially of wheat and rice in many developing countries during the 1970's and 1980's. However, in India, both the major food crops have failed to maintain the up trend of production. This fact shows that the yield levels have reached a plateau. Application of wrong technology or over-exploitation of natural resources may have short term gains but could often lead to ecological degradation, no longer sustainable by man. In fact, there are indications of decreasing production in areas where high external inputs have been consistently used over longer period (as is evident from results of several long term fertilizer experiments). In addition, adoption of such technology has created serious problems of salinity, water logging, soil erosion and degradation of environment including air and water pollution. This calls for re-orientation of research to develop suitable farming techniques for sustainable food production. The aim of sustainable farming system should be to produce enough to meet the demand and needs of the society, be profitable to the practitioner and conserve natural resource base and provide healthy and safe environment in the long run.

The inter-related, inter-dependent and inter-linking nature of IFS, involves the utilization of primary produce and secondary produce of one system as basic input of the other system, thus making them mutually integrated as one whole unit. This incidentally helps to reduce the dependence on procurement of inputs from open market, making thereby the IFS a self-supporting entity and sustainable system overtime.

3.3 Integration of subsystem in farming system

The various enterprises that could be included in the farming system are crops, dairy, poultry, goat rearing, fishery, sericulture, agro-forestry, horticulture, mushroom cultivation *etc*. It deals with utilization of wastes and residues. It may be possible to reach the same level of yield with proportionately less input in the integrated farming and the yield would be interestingly more sustainable because the waste of one enterprise becomes the input of another, leaving almost no waste to pollute the environment or to degrade the resource base. To put this concept into practice efficiently, it is necessary to study linkages and complementarities of different enterprises in various farming systems. The knowledge of linkages and complementarities will help to develop farming system (integrated farming) in which the waste of one enterprise is more efficiently used as inputs in another within the system (Fig. 2).

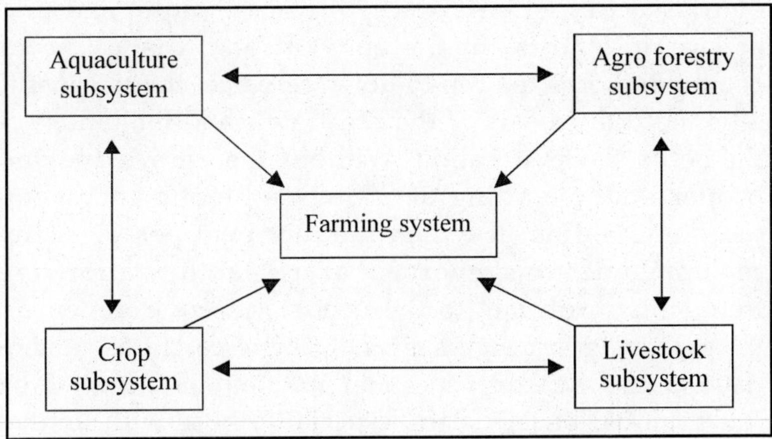

Fig. 2. Linkages of different enterprises in farming system

3.4 Goals of Integrated Farming System

The four primary goals of IFS are:

- Maximization of yield of all component enterprises to provide steady and stable income at higher levels.
- Rejuvenation/amelioration of system's productivity and achieve agro-ecological equilibrium.
- Control the build-up of insect-pests, diseases and weed population through natural cropping system management and keep them at low level of intensity.
- Reducing the use of chemical fertilizers and other harmful agro-chemicals and pesticides to provide pollution free, healthy produce and environment to the society, at large.

Thus, farming system as a concept takes into account the components of soil, water, crops, livestock, labour and other resources with the farm family at the center managing agriculture and related activities (Fig. 3). The farm family functions within the limitations of its capability and resources, the socio-cultural setting, and the interaction of these components with physical, biological and economic factors. In the conventional discipline or commodity-oriented top-to-down research, the scientist is often not fully aware of the constraints and limitations of the actual farm situation. Without the knowledge and proper understanding of the agro-climatic and socio-economic milieu in which the farmers operate, many recommendations evolved at research station are found inappropriate to the needs and environment of the practicing farmer. As a result, the research recommendations lose their relevance and are often not accepted by the potential beneficiaries, especially those with limited resources. There is, thus, a need for an alternate strategy to improve their efficiency through appropriate technology generation and diffusion.

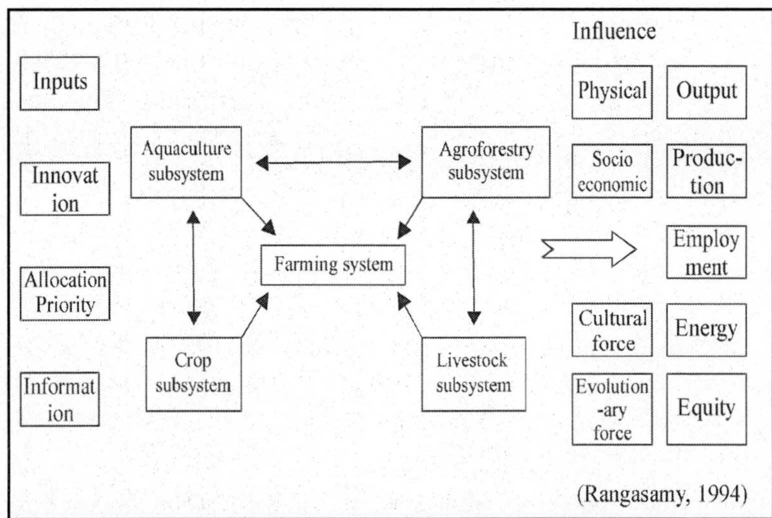

Fig. 3. Integration of subsystems into a sustainable farming system

3.5 Advantages of Integrated Farming System

a. Productivity

Integrated Farming System provides opportunity to increase economic yield per unit area per unit time by virtue of intensification of crop and allied enterprises. Time concept by crop intensification and space concept by building up of vertical dimension through crops and allied enterprises are the ways to increase productivity.

b. Profitability

The system as a whole provides opportunity to make use of the produce/waste material of one component as input on another component at the least cost. Thus, by reduction of cost of production of component two and extending similarly on other components linked, the profitability per rupee invested are being enhanced by eliminating the interference of middleman in most of the inputs used and working out the net income for the farm as a whole the benefit - cost ratio is increased.

c. Potentiality / Sustainability

Of late with an enthusiasm to produce more and more within the land area available to meet the requirement of additional population recorded at 2.2 % every year huge quantum of inorganic fertilizers, inorganic pesticides, fungicides, herbicides *etc.*, are dumped. Thus, there is every likelihood of soil and environment becoming polluted. Once when we loose larger area by virtue of the problem indicated, the productivity of the soil would automatically be reduced in the years to come. In IFS organic supplementation through effective utilization of by-products of linked components as manures is possible and thus it will certainly provide an opportunity to sustain the potentiality of the production base *viz.*, soil for much longer periods.

d. Balanced food

In IFS, components of varied nature are linked enabling to produce different sources of nutrition *viz.*, protein, carbohydrate, fat, minerals, and vitamins *etc.*, from the same unit area. This will provide an opportunity to solve the malnutrition problem that exists in the diet of the Indian farmers.

e. Pollution free environment

In the process of production in the crop based activity, some of the organics are left as waste material and such materials when ignored as such will pollute the environment on decomposition. Similarly, application of huge quantity of fertilizer, pesticide, fungicide and herbicide pollute the soil, water and environment to an alarming level. In the case of Integrated Farming System, waste materials are effectively utilized by linking appropriate components and thus utilizing the by-products as organic manures. Similarly by adopting bio control measures will help in preventing pollution of soil, water and atmosphere. In areas where piggery is a major activity, dumping of pig dung gives out huge quantity of unwanted gas and pollutes the atmosphere in the longer run. In IFS, the pig dung is used in the biogas unit to produce bio energy for the utilization of lighting, lifting water from the underground source and for running incubation unit of the poultry farm. Thus, IFS absolutely helps in avoiding environmental pollution.

f. Recycling

IFS establish its stability due to effective recycling of produces/waste materials of any one of the component as input on the other component linked in the programme. Thus, by way of recycling their own material at the farm level, the farmer could be able to reduce the cost of production which enables ultimately to increase the net income of the farm as a whole. Moreover, it also helps in reducing the environmental pollution expected out of decomposition of organic residues of the farm activity.

g. Money round the year

Unlike conventional crop activity where the money is expected only at the time of disposal of the economic produce received after five to fifteen months depending upon the duration of the crop, the IFS provides flow of money to the farmer round the year by way of disposal of eggs, milk, edible mushroom, honey, cocoons of silkworm, *etc.* This will help the resource poor farmer to get away from the clutches of usurious money lenders and from other financing agencies.

h. Adoption of new technology

Most of the big farmers are fully aware of the impact of new technologies included in the package. But, more than 80 per cent of the farmers who have been grouped under small and marginal categories are not able to execute the advanced technology proposed for want of money. But to IFS farmers, because of the linkage of Dairy/ Mushroom/ Sericulture/Fruit crops/Vegetables crops/Flower cultivation, *etc.,* money revolves round the year by sale of produce from these components at weekly/fortnightly interval. Availability of cash round the year gives a sort of inducement to the small and marginal farmers to go for the adoption of technologies like fertilizer application, pesticide and herbicide application, *etc.,* given in the package which otherwise is not possible under conventional farming due to paucity of funds.

i. Solve energy crisis

It is expected that the entire world is going to suffer for want of fossil fuel from 2030 AD. So it becomes inevitable to identify an alternative source to solve our energy crisis within a span of 3 to 4 decades. In IFS, by way of effective recycling techniques the organic wastes available in the system can be utilized to generate biogas. Though this may not be a source for complete supplementation, atleast to certain extent the energy crisis anticipated can be solved.

j. Solve fodder crisis

In IFS, each and every cm^2 of land area is effectively utilized. Growing of perennial fodder legume trees in the borders and water courses is a recommended practice in IFS. This practice not only helps in supplementing legume fodder but also enriches soil nutrient by fixing the atmospheric nitrogen. In the cropped land also IFS envisages intensification of cropping by including legume fodder like cowpea either as second tier or as third tier in the system. These practices will certainly relieve the crisis of non-availability of quality fodder to the animal component linked.

k. Solve fuel and timber crisis

The national demand of fuel wood in 2020 AD is 400 million m^3, where as the current production is only 20 million m^3. Similarly, the requirement of industrial wood in 2020 AD is 64.4 million m^3 and the current production level is just 11 million m^3. The present level of production should be increased to twenty folds in case of fuel wood and six folds in industrial wood. This could be possible to certain extent by afforestation programme in the shrub jungles and sparse forest areas. In IFS by linking agro-forestry appropriately, the production level of fuel wood and industrial wood can be enhanced without detrimental effect on crop activity in the field level.

l. Avoid degradation of forests

There is a vast gap between the demand and production level as far as fuel wood and timber are concerned. This will naturally induce the users to encroach on the forest area nearby illegally to bridge the gap. Right now our forest area is less (22%) than the prescribed norms (33%), to the geographical area available. When such encroachments are encouraged, there is every possibility of our forest area going to be a wasteland in the years to come. Even in the forest area available at present more than 2/3rd is sparse forest. The statistics on soil erosion indicates that every year 5374 million metric tones of built up soil is eroded in our country every year. This is four folds higher than the prescribed norms *viz.,* 4 mt ha^{-1} year^{-1}. By linking Agro-forestry in IFS, the degradation of forest area could be minimized to certain extent by supplementation of fuel and timber wood. By way of preserving the natural eco-system in the catchment areas, precious built up soil can also be preserved from erosion danger.

m. Employment Generation

Combining crop enterprises with that of livestock to take advantage of complementary and supplementary relationships between them, would increase the labour requirements tremendously and can help in solving the problem of underemployment to a great extent. IFS provide enough scope to employ family labour round the year.

n. Improves literacy

The farmers, who adopt IFS by combining different components like fishery, sericulture, mushroom cultivation, apiary, spawn production, dairy, poultry, agri-horticulture, agro-forestry, biogas production, *etc,* get expertise in each and every aspects of individual component on long range adoption. This will help the farmer to face any challenge in their activity.

o. Provides opportunity for agri-oriented industries

When once the produces of different components linked in IFS are increased to commercial level and if there is glut in the market, leads to the development of allied industries for preserving the by-products.

p. Increase input efficiency

The IFS provides enough scope to use the inputs on different components very effectively. This again leads to increased benefit - cost ratio.

q. Improves the standard of living of farmers

When once provisions are made in the farm level to generate bio energy, produce edible mushroom, fruits, eggs, milk, honey, vegetable, *etc.*, for the family use of the farmer apart from commercial purposes through IFS, create a sort of feeling among the farmers that they are no way inferior to other professionalists in the region. When they feel that their standard of living is on par with others and this will act as a booster tonic to continue agricultural profession without any reluctance which exists at present among most of the farmers.

3.6 Integrated Farming System for different Agro Climatic Zones

a. India

1. High altitude cold deserts

Pastures with forestry, goats, angore rabbits and a limited settled agricultural crops like millets, wheat, barley, and fodder crops.

2. Arid and desert region

Centering mainly in animal husbandry with the camels, sheep and goat and with moderate cropping components involving bajra, wheat, pulses, gram and fodder crops.

3. Western and Central Himalayas

Horticultural crops as a major component and have a less intensive agriculture mainly on the hill terraces and slopes with maize, rice, wheat, pulses and fodder crops.

4. Eastern Himalayas

Primitive crop husbandry with rice, millets, pulses, *etc.* Agro-forestry system are also common. Piggery and poultry are the chief livestock activity.

5. Indo-Gangetic plains

Intensive crop husbandry, involving rice wheat maize mustard-pulses and livestock inclusive of dairy cattle and buffaloes.

6. Central and Southern highlands

Cotton- sorghum millets-pulses with dairy cattle, sheep and goats and poultry are the secondary livestock and animal husbandry enterprises.

7. Western ghats

Major activity on plantation crops, cultivation of rice and pulses are the secondary agricultural activity. Cattle, sheep and goats are the livestock components which in most parts are maintained as large herds and allowed to range.

8. Delta and Coastal plains

Rice cultivation along with fish culture, poultry and piggery enterprises. Capture fisheries of the marine ecosystem is a specialized enterprise and does not mix with cropping activity.

(Shanmugasundaram, 1994)

b. Tamil Nadu

Name of the zone	Crop components	Other components
North Eastern zone	Rice, sugarcane, bajra. ragi, groundnut	One or two buffaloes or jersey cross breed cows or one unit of 20 goats or sheep
North Western zone	Ragi, groundnut, horsegram, rice, tapioca	Sheep or goat or poultry or dairy
Western zone	Millets, cotton and rice	Dairy, or poultry or sheep or goat

Contd...

Cauvery Delta zone	Rice	Dairy or goat or duck or piggery
Southern zone	Millets, cotton, pulses, rice	Sheep or goat or poultry or duck
High rainfall zone	Rice, tapioca	Dairy or duck
High altitude and hilly zone	Tea, coffee	Rabbit or goat or dairy

(Balasubramanian, 1994)

3.7 Components in integrated farming systems

Integrated farming systems comprising of crop and livestock has been sustainable over centuries. The system could be able to meet food needs of the ever increasing population till recently. In this system, animals are raised on agricultural waste and the animal power is used for agricultural operation end voids are used as manure and fuel. Tiwari (1993) reported that in the integrated farming the yield would be inherently more sustainable because the waste of one enterprise becomes the input of another leaving almost no waste to pollute the environment.

Integrated farming systems comprising the components of pig, duck, fish, azolla for the rice dominated La union of Philippines was suggested by Gavina (1990). However, earlier Balasubramanian (1988) identified sheep and buffalo, poultry and swine as dominant components in the rice regions of Philippines. Similarly, Singh *et al.* (1988) suggested the integrated farming systems comprising arable farming and mixed farming of buffalo, crossbred cows under irrigated condition and goat sheep rearing on grazing under drylands for Punjab region of India.

Predominance of milch cow, work bullock and vegetable farming along with cropping under garden lands in Western agro - climatic zone of Tamil Nadu was observed by Balasubramanian and Sankaran (1988). Whereas, Rangasamy *et al.* (1988) proposed poultry and fish components for Western zone wetlands of Tamil Nadu and

Chatterjee and Mandal (1988) suggested rice cum fish culture integrated farming systems with fruit crops for West Bengal region. Introduction of poultry cum fish culture in the rice cropping region of Thanjavur delta was suggested by Govindan and Chinnaswami (1988) and Govindan *et al.* (1990). Rice + fish farming for double crop wetland regions of Malaysia was suggested by Ali (1988) and Ali (1990) with various fish culture intensification techniques of concurrent, rotational and in between rice crops. Libunao (1990) suggested integration of aquaculture with livestock production for coastal provinces of Ilocos region of Philippines.

Rice + fish + azolla farming systems were developed for low lying wetlands of Western Zone (Shanmugasundaram and Balusamy, 1993) and for Thanjavur deltaic region by Shanmugasundaram and Ravi (1992) for upgrading the farm production potential through integrated farming systems. Similarly, rice + fish integration for small farm holding in Malawi in South Africa was recommended by Noble and Raheja (1990) in the lowland agro-eco systems. Whereas, Thanh Hung (1992) suggested fresh water prawn and marine shrimp culture for integration to coastal rice farming systems in Vietnam.

Introduction of an integrated rice + fish + vegetable farming system for deep water rice growing regions of West Bengal was successful to optimize the rice production and to sustain the eco systems (Roy *et al.*, 1990). However, while studying the feasibility of adopting aquaculture to increase the resource productivity in the existing Bangladesh farming systems, Ahmed and Rab (1992) suggested rice - cow - poultry - fishculture farming systems for small farm holdings. Tiwari (1993) while describing several crop-livestock farming systems in India, highlighted crop - livestock - agroforestry, flooded rice - poultry - fishery farming and rice - poultry - fishery - livestock system for resource rich farms and labour intensive farming systems like poultry, piggery, fishery, earthworm breeding and mushroom cultivation for small areas of high population, as dominant components in the farming systems.

Recent research information revealed that considerable fish production could be achieved by integrating fish with livestock where large amount of nutrient present in the feed could be recovered in the manure. Taiganides (1978) reported that further manuring the pond would enhance the plankton production including phyto, zoo and bacteria planktons which are high protein, natural food for protein supplement.

a. Duck cum fish farming systems

Recycling the waste of duck and goose on duck-fish and goose-fish integrated farming systems recorded higher productivity and return in the fish culture components in Hongkong (Sin, 1980). Similarly, Woyanarovich (1980) reported that duck voiding increased the common carp production in duck-fish integrated farming. Chen and Li (1980) reported that 5670 kg of annual fish production could be obtained from one hectare fish pond by using 2200 ducks voidings. According to Barah *et al.* (1982) polyculture with major Indian carps in an area of 400 m^2 fed with duck droppings showed daily gain of 38.5 kg fish yield per hectare. Sharma *et al.* (1985) reported that droppings from 300 ducks could be able to produce 3941 kg of fish ha^{-1} $year^{-1}$ in duck - fish integrated farming systems.

b. Pig cum fish farming systems

Polyculture of carps, catfish and large mouth fish with recycling of wastes and voids from 66 pigs as the only source of nutrients yielded a fish meat of 4 t ha^{-1} $year^{-1}$ in pig - fish integrated farming systems (Buck *et al.*, 1978). Jhingran and Sharma (1980) reported that voidings from 13 numbers of piglets produced 730 kg of fishes in 0.1 ha fish pond and Woyanarovich (1980) found that 60 pigs with fish combination gave the highest yield of 1950 kg fish ha^{-1} in integrated farming systems. According to Paul Raj (1984) composite fish culture by recycling pig dung resulted in one kg of fish production for every 10.4 to 11.6 kg of pig dung application.

c. Poultry cum fish farming systems

Integration of fish culture with poultry farming has recently been initiated in India and poultry manure could either be collected from poultry sheds and applied to fish ponds or poultry sheds be erected on the fish ponds. According to Ray and David (1969), most of the important soluble inorganic salts required for imparting productivity of water are present in large quantities in poultry droppings followed by cow dung and pig voids. A fish yield of 670 kg ha^{-1} in 90 days was recorded using poultry manure with no supplemental feed (Banerjee *et al.*, 1979). Similarly, Djajadiredja *et al.* (1980) reported that 650 kg of fish could be obtained from 2400 m^2 pond area with voidings of 120 poultry birds in Indonesia.

Sharma *et al.* (1985) reported that possibility of using fresh droppings from 300 layer birds for growing five species of carps each at the rate 800 fingerlings ha^{-1} in poultry cum fish farming systems. Kumar and Singh (1984) found that pellatizing poultry litter with 18 per cent protein content and feeding common carp fingerlings at four per cent body weight resulted in faster growth than with traditional feeding of groundnut cake and rice bran. Govindan (1988) obtained 4500 kg of fish yield ha^{-1} at Aduthurai by using fresh dropping from 500 layer birds with six species at a stocking density of 10000 fingerlings ha^{-1} in poultry-fish integrated farming systems.

d. Cropping cum mushroom farming system

Besides livestock integration with cropping, mushroom cultivation could also be a better proposition in integrated farming system. Rao (1984) opined that villagers and farmers could take mushroom cultivation to their houses and farm as a component occupation in the farming to derive nutritious food and additional income. According to Vangriensven (1987) modern mushroom cultivation was one of the components of specialised farming in Netherlands which contributes larger parts of the total mushroom production.

Paddy straw substrate was found to be superior to wheat straw and more economical for mushroom production in farming systems (Jitendra Pal and Thapa, 1979). Vishwanath (1982) and Garcha *et al.* (1984) evaluated the bio-efficiency of paddy straw for higher mushroom productivity in rice-mushroom integrated farming systems. Loganathan (1989) suggested mushroom cultivation as a component in the integrated farming systems with rice dominated cropping region of Thanjavur. Similarly, Rangasamy and Jayanthi (1994) suggested integration of mushroom cultivation with fishery - poultry in rice based cropping system dominated lowlands of Western zone of Tamil Nadu.

e. Agro-forestry in farming system

Trees are an indispensable constituent in the farming system. The vital role of forests and trees in serving the mankind by providing food, fibre and shelter, the three basic necessities of life. In most parts of the country rural people recognize the role of trees in providing a number of locally important goods and services. The "Alley farming" is becoming popular and gaining momentum which is an agro-forestry system where livestock production is integrated into agricultural and forestry crops / trees. They commonly exhibit a layered vertical structure of trees, shrubs and ground cover plants that recreate some of the properties of nutrient recycling, soiled protection and effective use of space above and below the soil surface. These are widely used to supplement outputs from other parts of the farm. They are also used to spread farm work, output and income more evenly throughout the year.

Role of tree components in farming system

Tree growing practices contribute to a wide variety of existing farming system. They tend to be prominent parts of systems where capital and physical resources are limited. In such conditions a tree component within the farming system can play one or more of the five following overlapping roles:

- Make productivity of land in situation of scarce capital through the substitution of trees for purchasing inputs like fertilizer, herbicide and investments in soil, crop protection and / or irrigation water.
- Make productive use trees, as low-input, low-management crops, constitute the most effective use of these resources.
- Maximize returns to land in situations where land and capital are limited, and tree / crop livestock combinations permit fuller use of available labour than alternative use of land.
- Increase income earning opportunities from use of farm resources as size of land holding and / or site productivity fall below the level at which the household basic food needs can be met from on-farm production of food.
- Strength risk management through diversification of outputs, wider seasonal spread of inputs and outputs and reduction in the likelihood of crop failure through drought, and build-up of tree stocks to provide capital.

(Balakrishnan, 1995)

f. Multipurpose trees in farming system

The term refers to trees which make more than one major contribution of the land use systems in which they are grown. Most of the trees used in farming system grow relatively quick, while many also fix nitrogen through symbiotic association with microbes. Multipurpose trees have productive and service functions. The leading products are the 3 F's, fuel wood, fodder and fruit. Other uses are timber, poles, nuts, oils, beverages, fibres, thatching and medical substances. The service functions include shade from the sun for animals and men, conservation of soil fertility, fencing and water conservation.

Trees improve soils by many processes, the most important of which are organic matter maintenance, nitrogen fixation, nutrient recycling and augmentation over grasses. They give high-protein fodder, and as a result of rooting systems which reach down to the water table, remain green at times of the year when grasses have died back and become unpalatable. Fodder banks are woodlots, consisting of high protein, fodder producing trees such as *Leucaena leucocephala*. They can be employed as a component in stall-fed cattle rearing or to provide supplementary feed to grazed cattle. A leguminous species, *Acacia senegal*, improves soil fertility as well as produces gum arabic for sale, and fuel wood, medicine, fibres and other products for use in the household.

3.8 Contribution by components in integrated farming system

Arable farming only ends in narrow profit or less margin to the farmers, warranting development and adoption of integrated farming systems involving various possible combinations of components. Reheja and Oberai (1953) and subsequently Singh (1957) considered that, integrated farming in India implied the combining of cropping and livestock production for efficient and effective uses of resources to the best advantage of the farmers. Consequently, Arora (1980) opined that integrated farming systems as the complementary symbiotic exploitation of crop and livestock to get maximum advantage.

a. Economic contribution

According to Muthiah (1961) and Rajagopalan *et al.* (1961), the percentage increase of returns in integrated farming systems was steady and higher than cropping alone. About 90 per cent of additional income was reported with integrated farming systems by Talathi and Borude (1980) over arable farming. Integration of milch cow with cropping resulted in higher net returns through maintaining high yield of crop and animal than cropping alone (Ravi Varma, 1975, Sundaresan, 1975). Investigation by Bhattacharya (1976) and Vikaraman Nair *et al.* (1976) revealed that integrated farming, if properly practised would be more profitable than arable farming in small holdings. Sundaresan (1979) reported that livestock keeping was more suited to small land holders to fetch additional income to the farm family.

The economics of different types of integrated farming systems under marginal farming situation were evaluated by Chinnaswami *et al.* (1980) who reported that mixed farming practice linking dairy gave the highest return than pure cropping. Sirohi *et al.* (1980) found that there was an increase in the return to the tune of 148 per cent due to the introduction of poultry cum dairy based integrated farming systems over cropping. Cropping and dairy enterprises with crossbred cows fetch additional gross and net returns than cropping (Singh and Sharma, 1983), while comparing integrated farming systems with cropping alone under rainfed conditions. Mohamed Ali *et al.* (1984) opined that dairy cum poultry based integrated farming systems were the best to get higher net annual income. Assured net return could be obtained with duck rearing in an integrated farming system for the investment made in the farming systems (Sharma *et al.*, 1985). Govindan (1988) found that the income generation with integrated farming systems with poultry cum fish culture was higher than in rice based cropping.

Higher net income with integrated farming systems comprising cropping-poultry-fish-mushroom culture than with rice based cropping system under lowland condition was recorded by Rangasamy *et al.* (1990). They also recorded additional income through integration of dairy - biogas - silviculture over cropping under garden land condition and cropping - goat farming over conventional cropping under rainfed situation in the Western zone of Tamil Nadu. In Cauvery delta regions of Tamil Nadu, higher net returns could be generated by integrated farming systems with poultry cum fish culture as components (Govindan *et al.*, 1990). Siddeswaran and Palaniappan (1990) reported that added income could be realized through dairy based integrated farming systems under garden land conditions of Western zone of Tamil Nadu. Balusamy and Shanmugasundaram (1994) and Balasubramanian (1994) registered maximum net return and per day profit with rice - fish - azolla integrated farming systems as compared to conventional cropping under lowland condition.

Crop + pigeon + fish + mushroom integration was found to be superior in obtaining highest net returns with higher per day return and benefit cost ratio in marginal lowland farming situation of Western zone of Tamil Nadu (Jayanthi *et al.*, 1997). Balusamy (1996) reported that integration of rice - rice + Azolla + fish farming system with 75 per cent recommended N supply as well as incorporation of green leaf manure resulted in higher productivity with increased economic return.

Cropping + dairy + biogas + mushroom + fish (cattle shed washing) integration earned the highest returns with better per day return for irrigated uplands of Coimbatore was reported by Sivamurugan (2001). For rainfed farming, Radhamani (2001) reported that integration of crop + tree + goat system provided higher net return than cropping alone for Western zone of Tamil Nadu. Jayanthi *et al.* (2002) quoted that integration of goat and fish with cropping nourished with recycled fish pond silt with goat droppings gave higher net return and per day return for wetland ecosystems of Western zone of Tamil Nadu.

b. Employment generation

Integrated farming systems have become more relevant as they provide increased opportunities for employment among the rural poor family (Swaminathan, 1975). Ravi Varma (1975) quoted the results of the trials at Kasargode that one hectare of coconut gardening required 150 man days and it was increased to 1000 man days on introduction of dairy based integrated farming. Sundaresan (1979) reported that livestock keeping is a labour intensive effort with the involvement of labour round the year and Gill (1979) expressed that integrated farming systems could help in efficient utilization of farm family labour.

Cropping alone generated 400 man days as against 904 in integrated farming systems with six buffalo (Pandey and Bhogal, 1980). Talathi and Borude (1980) reported that integrated farming systems provided more gainful employment than that of arable farming. Chinnaswami (1981) reported that cropping alone generated 303 man days and integrated farming systems with dairy cum poultry and dairy alone generated 530 and 513 man days, respectively. Rao (1983) reported that cattle and buffalo rearing involved intensive use of family labour and offered significant employment opportunity for small and marginal farmers. Narinder Singh *et al.* (1984) stated that a herd of 200 goats under integrated farming systems provided full time employment for two persons throughout the year.

Labour utilization was found to increase by 182 per cent in integrated farming systems by the introduction of 270 poultry birds in a crop cum poultry enterprise (Sirohi *et al.*, 1980). According to Mohamed Ali *et al.* (1984) employment generation to the tune of 550 man days was possible in dairy cum poultry based integrated farming systems. Cropping alone generated 245 man days and integrated farming systems with sericulture in one hectare generated 598 man days in a year (Satheesh *et al.*, 1985). Gupta and Patel (1985) reported that employment generation through integrated farming systems involving 30 milch cows was 3058 man days in a year.

Additional employment of 124 man days with poultry cum fish linked integrated farming systems than with cropping alone was reported by Govindan (1988). Maintenance of four milch cows with cropping could generate an additional employment of 274 man days as against cropping alone in Thanjavur delta (Chelliah and Chandrasekaran, 1989). Rangasamy *et al.* (1992) reported an additional employment of 184 man days in poultry - fish - mushroom integrated farming systems over conventional cropping system under lowland farming. Premsekhar and Rangasamy (1994) stated that integration of crop - dairy - biogas - silviculture - spawn production could generate an additional employment of 562 man days than cropping alone under lift irrigated garden lands. Chandrasekaran *et al.* (1994) inferred that integration of duck cum fish culture and dairying could generate 396 and 702 man days, respectively, as against 252 man days with rice based cropping alone in Cauvery delta region of Tamil Nadu.

Jayanthi (1995) reported that integration of cropping with poultry + fish + mushroom generated the highest employment of 798 mandays. The allied enterprises added employment to the tune of 423 mandays providing opportunity for 1.16 family members to be employed per day round the year. Sivamurugan (2001) stated that integration of cropping with dairy + biogas + mushroom generated the highest employment of 875 mandays. According to Radhamani (2001), the additional employment gained through integrated farming system with crop + goat was 314 mandays ha^{-1} $year^{-1}$ under rainfed *vertisols*. For lowlands, integration of crop + fish + goat gave 575 mandays of employment ha^{-1} $year^{-1}$ (Jayanthi *et al.*, 2002).

c. Resource contribution

Integrated farming systems involving dairy cattle became more relevant under existing ecological and socio-economic condition in India and would provide an increased quantity of manure for cropping (Swaminathan, 1975). Vikaraman Nair *et al.* (1976) noted that there was little loss of nutrient form the field as those fodder crops removed were returned in the form of cattle manure in integrated farming systems. Sundaresan (1979) stated that livestock keeping along with crop production would provide manure and nutrient, necessary for crop production.

Livestock could make a significant contribution to crop production through provision of power and manure which were of growing concerns as sources of energy and potential substitute of fossil fuel (Harbovszky, 1981). Rao (1983) observed that cattle and buffalo were the main sources of draught power for agriculture operation and would provide large quantities of animal byproducts as valuable organic manures. Mohamed Ali *et al.* (1984) stated that by recycling the wastes of dairy cow and poultry birds, the value of nutrients added to the soil could be increased.

Crop-livestock integrated farming systems would quite be sustainable because the resources are recycled which help in reducing the external input and increasing that soil productivity. Rice - poultry - fishery farming systems also increased the nutrient uptake by rice while increasing the organic matter content of soil (Tiwari, 1993). Integration of rice - fish culture also helped in improving the soil fertility and enhancing the resource recycling efficiency among the enterprises in the system (Lightfoot *et al.*, 1993). Mishra and Pasupalak (1994) stated that in intensive integrated farming systems involving duckery cum fish culture provided excellent opportunity to organic recycling and reduced farmers dependency on market purchased inputs.

Integration of enterprises like cattle rearing, fishery, poultry and goat rearing, sericulture and mushroom cultivation with cropping could properly recycle the residues for deriving maximum compatibility and replenishment of organic matter. Fish integrated farming was ideal for better utilization for farm resources and conservation of the eco-system in rainfed lowland farming.

Replenishment of soil fertility status through substantial improvement in the post harvest available NPK nutrients could be achieved even with higher removal of nutrients through crop uptake by the application of recycled or composted pigeon and poultry manure combined with inorganic fertilizer to each crop in the cropping system was reported by Jayanthi *et al.* (1997). Balusamy (1996) opined that rearing of fish in rice field resulted in higher soil nutrient status at the end of the cropping period and the soil available N was not depleted much, even after heavy uptake through crop removal in rice-fish culture systems.

Application of 50 per cent nitrogen through fertilizer and 50 per cent through goat manure enhanced the soil fertility status and provided better opportunity for recycling of manure to the crops was observed by Radhamani (2001) under *vertisols* of Western zone of Tamil Nadu.

3.9 Crop and animal waste utilization

The productivity of livestock mainly depends on the availability of quality feed and fodders in requisite quantity.

It is observed that most of the livestock population in the country is underfed. The residues of different agricultural crops constitute the major source of fodder for livestock. In India, till late 1980's the main source of livestock feed consisted of conventional agricultural residues like rice straw, wheat, sorghum and maize stalks, sugarcane trash and remains from pulses. Based on the grain: straw ratio, approximately 321.4 million tonnes of agricultural crop residues are available in India.

The area under green fodder is not increasing to meet the increased demand. The area under green fodder in the country is estimated to be approximately 4% of the total cultivated area. Since the farmers feel that animal power is one of the costliest inputs, they switched over to mechanical power for most of the operations. Moreover, the animals meant for milch and meat is replaced by cross-bred and improved varieties, which need to be supplemented with quality concentrated feeds. Thus, the estimated total production of crop residues in 2020 AD to the tune of 447.0 million tonnes is to be effectively utilised otherwise. The details of the crop residues in India as given by FAO (1985) projected to 2020 AD are furnished in Table.1.

In addition to crop residues, there is a possibility for collecting the following quantities of animal voids in India (FAO, 1985).

Cow dung	750 m.t.
Buffalo dung	250 m.t.
Voids of small ruminants	130 m.t.
Total	1130 m.t.

At present the voids are being utilised for fuel and as FYM/compost. The same objectives can very well be attained with boosted value through recycling and as identified through experimentation. When recycled through biogas unit, there is good possibility to improve the organic source of nutrients apart from generation of fuel energy when tapping of fossil fuel is getting depleted. Apart from the major nutrients there is good amount of enhancement in the availability of secondary and trace elements required for the excellent performance of crop commodities. By virtue of adoption of one of the modern agricultural technology *viz.*, Integrated Farming System (IFS) in the farm activity, there is a possibility of improving untapped potential of each and every produce by recycling with dual benefits.

Table 1. Projected availability of crop residues in 2020 AD

Crop	Quantity of residue (million tonnes)	
	FAO estimate (1985)	Projected for 2020 AD
Rice	118.9	160.0
Wheat	57.5	80.0
Maize	21.0	35.0
Millets	40.0	56.0
Sorghum	41.0	52.0
Sugarcane	43.0	64.0
Total	**321.4**	**447.0**

The schematic diagram projecting part of animal voids for generation of bio-energy and utilization of slurry as organic source of nutrients as well as availability of crop residue and recycling for mushroom production and their nutrient value are furnished in Fig. 4 and 5.

3.10 Enterprises of Farming System

In the integrated farming system, it is always emphasized to combine cropping with other enterprises/ activities. Many enterprises are available and these include cattle maintenance, sheep or goat rearing, poultry, rabbit

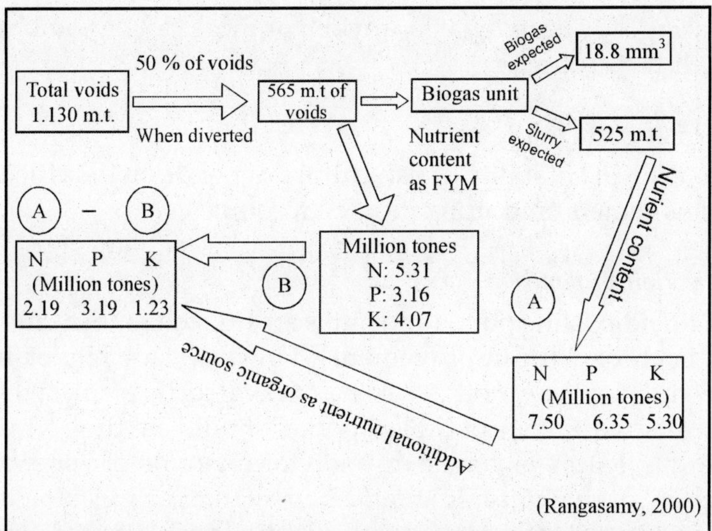

Fig. 4. Projected animal voids and recycling for biogas production with its nutrients in India

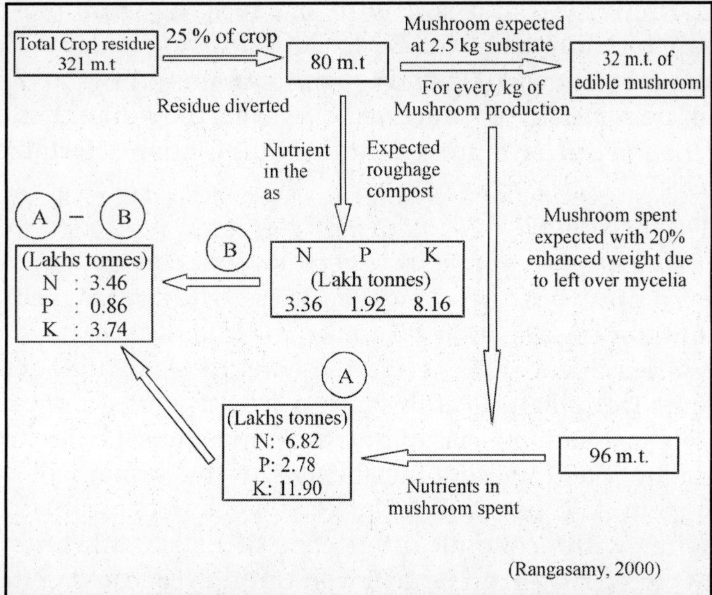

Fig. 5. Projected crop residue and recycling for mushroom production with its nutrients in India

rearing, bee keeping, etc., anyone or more can be combined with the cropping.

a. Fish culture practices

The profitability of fish culture depends on the efficient management and utilization of resources.

Selection of pond

The existing ponds and tanks are utilized for fish culture as such or with improvements. However, for higher fish production new ponds can be excavated in places where ideal conditions exist. Fish ponds should be dug on non porous, loamy or clay loamy soil for retention of water and better productivity. It should be rectangular in shape and of easily manageable size (0.1-0.4 ha) with assured water supply for replenishment as and when required. A depth of 2 m is considered congenial for good biological productivity. Pond soil with pH ranging from 6.6-7.5, available nitrogen from 30 to 50mg/100g, available P_2O_5 from 6 to 15 mg/100g and organic carbon from 1.0 to 2.5% will be naturally productive. However, the nutrient status of the soil can be improved through application of fertilizers.

Pond preparation

Occurrence of macro vegetation will consume a lot of nutrients from the pond sediment and water, reduce plankton production and make it difficult to exploit the fisheries. Hence, it is advantageous to clear the aquatic plants from the ponds through mechanical or chemical or biological methods. Similarly, the ponds must be free from predatory and uneconomical fishes, as the former will prey upon stocked fish and the latter will compete for food and shelter with economical varieties of fish. Eradication of unwanted fishes can be achieved through repeated netting, draining the water and exposing the pond to sunlight and application of toxicants to the pond.

Before stocking pond should be cleaned (aquatic vegetation, predator, weeds and excessive silt). The methods recommended are

- To clear dense vegetation, apply 2, 4-D Sodium salt
- Release of adequate numbers of Grass carp to control variety of weeds such as duck weed, Hydrilla, grass weeds
- Dewatering the pond. This is possible during summer months.
- Pond embankments should be strengthened to prevent entry to predatory fishes.
- Mahua oil cake to clear unwanted fishes. The toxicity remains for 15 days.

After 15 days, release of fingerlings should be done.

Pond fertilization

Organic and inorganic nutrient requirement

Mahua oil cake	:	200 kg / ha
Cow dung	:	10000 kg/ha
Super phosphate	:	250 kg/ha
Quick lime	:	500 kg/ha

The nutrient status for the actual pond area (10 cents)

Stocking the pond

A pond is a three dimensional growing space. To treat it like a field, by planting only one kind of crop, is likely to result in wasting major portion of the space. A fertile pond produces a number of fish food organisms (phytoplankton, zooplankton, benthos and epiphytic algae) in addition to organic detritus. Since, most of the fishes are not omnivorous, but rather selective in their diet, stocking of single species leads to wastage of not only space but also food. Food production can greatly be increased through polyculture of various species with different feeding habits. Polyculture allows full utilization of all the available space in the pond. It can increase the stocking density and fish output. Beneficial interactions between compatible species cultured in the same pond become evident in polyculture where all the species are mutually benefited and the production of each species is increased. Through the beneficial interactions between species, one kind of food can be doubly utilized. It is stated that in polyculture, one grass carp can provide enough food for three silver carp through proliferating natural organisms. Hence, in a polyculture system, healthy fingerlings (100-150 mm long) of Indian major carps (*C. catla, L. rohita* and *C. mrigala*) and exotic carps (*H. molitrix, C. idellus* and *C. carpio*) may be stocked in the following species ratio: surface feeders 30-40% (*C. catla* 10-15% + *H. molitrix* 20-30%); column feeders (*L. rohita* 15-20% in deep ponds or 10% in shallow ponds); bottom feeders 40-45% (*C. mrigala* 15-20% + *C. carpio* 20-25%) and macrophyte feeder (*C. idellus* 5-15%). Optimum stocking density is determined on the basis of local conditions to obtain a good harvest. Normally, it ranges from 5000 to 20000 fingerlings/ha depending upon the technology and management measures adopted.

Stocking of fingerlings: 10,000/ha (local carp and exotic carp alone)

S.No	Species	Mobility/ Feeding	Stocking density (No)	Stocking density (%)
1.	Silver carp (*Hypohthalmichthys molitrix*)	Surface	80	20
2.	Catla (*catla*)	Surface	80	20
3.	Rohu (*Labeo rohita*)	Column	80	20
4.	Mirgal (*Cyprinus mrigala*)	Bottom	60	15
5.	Common carp (*Cyprinus carpio*)	Bottom	60	15
6.	Grass carp (*Clenopharyngodon idellus*)	Grass	40	10

Supplementary feeding

The natural food resource of the pond is not adequate to meet the demand of the fish stock. Hence, supplementary feed must be provided for better growth and survival. The expenditure on feeds normally accounts for nearly 60% of the total production cost. The cost is greatly reduced in integrated farming systems where the waste product of one culture activity is utilized in another farming system.

Rice bran	:	40%
Maize flour	:	40%
Oil cake	:	20%
	(or)	
Rice bran	:	50%
Oil cake	:	50%

b. Cattle maintenance

Rearing of cattle in India is carried out under a variety of adverse climate and environmental conditions. The cattle are broadly classified into three groups.

1. *Draft breeds:* The bullocks of these breeds are good draft animals, but the cows are poor milkers. *e.g.* Nagore, Hallikar, Kangeyam, Mali.

2. *Dairy breeds:* The cows are high milk yielder but the bullocks are of poor draft quality *e.g.,* Sahiwal, sindhi, Gir

3. *Dual purpose:* The cows are fairly good milkers and the bullocks are work capacity. *e.g.,* Hariana, Ongole and Kankrej

Exotic breeds: The exotic breeds are high milk yielder, *e.g.,* Jersey, Holstein - Freiesian Aryshire, Brown Swiss and Guernsey.

Buffaloes

Important dairy breeds of buffalo are Murrah, Nili Ravi (presently in Pakistan), Mehsana, Suti, Zafarabadi, Godavari and Bhadawari. Of these Godavari has been evolved through crossing coastal buffaloes in coastal regions of Andh, Murrah.

Housing: Each cow required 12 to 18 sq. m space and the buffaloes need 12- 15 sq.m. It is important to provide good ventilation and an open shed of housing is always preferable. Dairy building should be located at an elevated place to facilitate easy drainage. The floor should be rough and have gradient of 2.5cm for every 25cm length.

Breeding and maintenance: The cows remain in milk for 9-10 months, the average calving interval being 16-18 months. A cow does not require more than 6-8 weeks of dry period.

From the economic point of view, cow should ordinarily be bred during the second and third months after calving. In weak animals and heavy milkers, 1 or 2 months may delay breeding. Cattle animals come to heat in more or less regular cycles of about 21 days, which lasts for about a day. The best time to serve a cow is during the last stage of heat. If artificially, inseminated, a second insemination may be given within 6-8 hours after the first insemination. The gestation period varies with individual cows and breeds and normally it about 280 days.

In the case of buffaloes, the lactation period last for 7-9 months. She-buffalo comes to heat every 21-23 days. The gestation period is 310 days. Calf rearing is very important in the case of buffalo maintenance. Since they require abundant water, wallowing is required. Regular deworming is needed for buffalo maintenance.

Under Indian condition, cattle commonly mature at the age of about three years. This period can, however, be reduced by six months under well managed herd.

Feeding: Cattle feed generally, contains fibrous, coarse, low nutrient straw material, called roughage and concentrates.

Roughages: Dairy cattle are efficient user of the roughages and convert large quantities of relatively inexpensive roughage into milk. Roughages are basic for cattle ration and include leguumes, non-legume hays, straw and silage of legume and grasses.

Concentrates: Grains and byproducts of grains and oilseeds constitute the concentrates. They are extensively used in dairy cattle ration. These include cereals (maize, sorghum, oats and barley), cotton seeds, industrial waste (bran of wheat, rice, gram husk) and cakes of oilseeds (groundnut, sesame, rape-seed, soybean, linseed).

Vitamins and mineral mixtures: It is advisable to feed a supplement containing vitamins A and D. Mineral mixtures containing salt, Ca and P should also be provided in the ration.

The ration per animal per day normally includes concentrate @ 1 kg for 2 litres of milk yield green fodder (20-30kg), straw (5-7kg) and water (32 litres).

c. Sheep and goat rearing

The system of sheep and goat rearing in India is different from that adopted in the developed countries. In India, smaller units are mostly maintained. In the developed countries, however, it is practiced in large scale in fenced areas.

However, activity of goat rearing is sustained in different kinds of environments, including dry, hot, wet and cold, high mountains or low lying plains. The activity is also associated with different systems such as crop or animal-based, pastoral or sedentary, single animal or herd, small or large scale. Goat is mainly reared for meat, milk, hide and skin. Goat meat is the preferred meat in India. A goat on hoof (live goat) fetches a better price than a sheep on hoof.

Housing: Goats can be maintained under stall-fed conditions. Successful goat rearing depends on the selection of site. Goats do not thrive on marshy or swampy ground. Goats are to be provided with a dry, comfortable, safe and secure place, free from worms and affording protection from excessive heat and inclement weather. The kids are kept under large inverted baskets until they are old enough to run along with their mothers. Males and females are generally kept together. The space requirement for a goat is 4.5 to 5.4 sq.m.

Breeding and maintenance: Goat matures in about 6-7 month. Breeding is allowed for buck at one year and doe after 10 months of age. Gestation period is 145-155 days. It gives birth to 1-3 kids per time. Numbers of evings are three per 2 years. The kids can be weaned after 30-40 days. Mother can be allowed for mating 45-60 days after eveing. Once in five years, the buck can be changed to avoid deterioration due to inbreeding. When the young ones attain a body weight of about 25-30 kg in about nine months, they can be sold.

Feeding: The per head requirement of nutrients in respect of goats is relatively low. Hence, they are suitable for resource poor small farmers with marginal grazing lands. They are essentially browsers and eat plants, which any other animals won't touch. Goats eat 4-5 time of their body weight. Since the profit depends on weight addition, adequate proteins and calorie should be giving to goats. They eat more of tree fodder and hence 40-50% of green fodder should contain tree leaf fodder the rest with other grass species. Goats should be fed with concentrates of maize, horse gram, groundnut cake, and fish meal and wheat bran. Common salt and vitamin mixtures should also be added. Abundant clean fresh water should be made available to the goats. Water should be changed every morning and evening. Fresh water is required for digestion, blood circulation and removal of waste form the body. Water is also required for regulation of the body temperature.

Sheep rearing: Sheep are well adapted to many areas, excellent cleaners and make use of much of the waste feed. They consume large quantities of roughage, converting a relatively cheap food into a good cash product. Housing need not be elaborate or expensive. However, to protect the flock from predatory animals, the height of the fencing should be raised to two metres.

Breeds of Indian sheep: There are three types of sheep in India based on the geographical division of the country.

1. *The temperate Himalayan region:* Gurez, Karanah, Bhakarwal, Gaddi, Rampur Bushiar.
2. *Dry western region:* Lohi, Bikaneri, Marwari, Kutchi, Kathiawari
3. *Southern region:* Deccani, Nellore, Bellary, Mandya, Bandur.

Breeding and maintenance: One ram can be maintained for 40-50 ewes. Rams are liable to fight when two or more of them are put in a pen. Unlike other farm animals, ewes in general do not come in heat at regular intervals throughout the year but are seasonal in this respect. The duration of the heat period will range from 1-3 days and 75% of the ewes remain in heat for 21-39 hours. The optimum time of service is towards the end of heat period. Average heat interval is 18 days during the breeding season. The gestation period will vary from 142-152 days with an average of 147 days. A normal ram is in full vigour for breeding from the age of 2 ½ -5 years. Sheep grow fully at two years of age when the ewe is ready for breeding. Under average range condition, may be expected to produce about five crops of lambs.

Feeding: A sheep requires about 1-2 kg of leguminous hay per day depending on the age of sheep and its body weight. Proteins may be supplied through concentrates such as groundnut cake, sesame cake or safflower cake when the pastures are poor in legumes or when scarcity conditions prevail. Normally 110-225 g of cake is sufficient to maintain an average sheep in good condition. Feeding a mixture of common salt, ground limestone and sterilized bone meal in equal parts is required to alleviate deficiency of minerals in the feed.

d.Poultry rearing

Poultry is one of the fastest growing food industries in the world. Poultry meat accounts for about 27% of the total meat consumed world wide, and its consumption is growing at an average of 5% annually. Poultry industry in India is relatively a new agricultural industry. Till 1950, it was considered a back yard profession in India. In the sixties, the growth rate of egg production was about 10% and it increased to 25% in the seventies. The growth rate came down to 7-8% by 1990 due to price-rise in poultry feed. By 2000, the total egg production may reach upto 5000 crores. Broiler production is increasing at the rate of 15% per year. It was 31 million in 1981 and increased up to 300 million in 1995 (Singh, 1997). Nearly 330 thousand tones of broiler meat are currently produced. The average global consumption is eggs per person per year and in India, it is only 32-33 eggs per capita year. As per the nutritional recommendation, the per capita consumption is estimated at 180 eggs/year and 9 kg meat /year.

Breeds: Specific poultry stocks for egg and broiler production are available. A majority of the stocks used for egg production are crosses involving the strains or inbred lines of white Leghorn. To a limited extent, other breeds like Rhode Island red, California Grey and Australop are used. Heavy breeds such as white Plymouth Rock, white Cornish and New Hampshire are used for cross-bred broiler chickens. Hence, it is essential to consider the strain within the breed at the time of purchase. Several commercial poultry breeders are selling day old chicks in India. It is best to start with the day old chicks.

Housing: Adequate space should be provided for the birds. Floor area of about 0.3-0.4 m^2 bird is required for heavy breeds. The house should have good ventilation and reasonably cool in summer and warm during winter. It should be located on well-drained ground, safe from flood waters.

Feed: The feed conversion efficiency of the bird is far superior to other animals. About 60- 70 per cent of the total expenditure on poultry farming is spent on the poultry feed. Hence, use of cheap and efficient ration will give maximum profit. Ration should be balanced containing carbohydrates, fats, minerals and vitamins. Some of the common feed stuffs used for making poultry ration in India are:

Cereals: Maize, barley, oats, wheat, pearl millet, sorghum, rice-broken.

Cakes/meal: Oil cakes maize -gluten-meal, meat meal, blood meal.

Minerals/salt: Limestone, Oyster shell, salt, manganese.

From the day old to 4 weeks of age, birds are fed on starter ration and thereafter finisher ration, which contains more energy and 18-20% protein. Feed may be given 2-3 times. In addition to the food stuffs, antibiotics and drugs may also be added to the poultry ration. Laying hens are provided with oyster shell or ground limestone. Riboflavin is particularly needed.

Maintenance: The chicks must be vaccinated against Ranikhet diseases with FI strain vaccine within the first 6-7 days of age. One drop of vaccine may be administered in the eye and nostril. When chicks get the optimum body weight of 1.0-1.5 kg around six weeks, they can be marketed for broiler. Hens may be retained for one year for production i.e., upto the age of about 1½ years. After that they are disposed off for table purpose. It may not be economical to keep the hens beyond 1 ½ years since egg production would get reduced. One hen is capable of laying 180-230 eggs in a year starting fro sixth month. In addition, a laying hen produces about 230 g of fresh droppings (75% moisture) daily.

e. Duck rearing

Ducks account for about 7% of the poultry population in India. They are popular in coastal and water logged states like West Bengal, Orissa, Andhra Pradesh, Tamil Nadu Tripura and Jammu and Kashmir. Ducks are predominantly of indigenous type and reared for egg production on natural foraging. Ducks are quite hardy, more easily brooded and resistant to common avian diseases. In places like marshy river side, wetland barren moors where chicken or any other type of stock do not flourish, duck farming can be a better alternative.

Breeds: The important Indian breeds are Sylhet Mete and Nageswari which are mostly found in the Eastern region of the country. Their annual production is 150 eggs/bird/year. Improved breeds for egg and meat production are available. Khaki Campbell and Indian Runner are the most popular breeds for egg laying. Khaki Campbell has a production potential of 300 eggs/bird/year. Indian Runner is the second best producer. White Pekin, Muscovy and Aylesbury are known for meat production. White Pekin is the most popular duck in the world. It is fast growing and has low feed consumption with fine quality of meat. It attains about 3 kg of body weight in 40 days. Indigenous types, however, still continue to dominate in duck farming. Desi ducks are robust, well adapted to local conditions and free of diseases.

Housing: Ducks prefer to stay outside day and night even during winter or rains. In the mild climate, it is possible to raise ducks without artificial shelter. A light fence of at least 1.2 m high enclosing the yard is enough to stop any predators. One nest of size 0.3 x 0.3 x 0.45 m to every 3 ducks is sufficient. In case of laying birds, a mating ration of one drake: 6-7 ducks and in meat type 1:4-5 is allowed. The duck house should be well ventilated, dry, leak and rat proof. The roof may be of thatched or asbestos sheeted. A water channel of 0.5 m wide and 0.20 m deep is constructed at the far end on both sides parallel to the night shelter in the rearing or layer house.

Feeding: Ducks normally require lesser attention. They supplement their feed by foraging, eating fallen grains in harvested paddy fields, small fishes and other aquatic materials in lakes and ponds. However, for intensive rearing, pellet feeding may be given. Ducks prefer wet mash due to difficulties in swallowing the dry mash. Hence, ducks should never have access to feed without water. During the first 8 weeks, birds should always have an access to feed. Later on they may be fed twice a day in the morning and late afternoon.

Maintenance: The general management of ducks is similar to that of the chickens. The incubation period is 28 days. A broody duck or hen may be used for small scale hatching and incubator for large scale hatching. During the early part of the life, newly hatched ducklings require warm temperature under the natural or village conditions. A duck or broody hen can take care of 10-15 ducklings. Artificial brooding may be resorted for large number of ducklings. High egg-laying strains of duck come into production at 16-18 weeks of age.

Ducks are resistant to common avian diseases. Some of the common diseases in duck are plague, duck plague, duck virus hepatitis, duck cholera and aflatoxicosis.

f. Turkey rearing

Turkey is a robust bird and be reared in humid tropics. It actively feeds on a variety of crop residues and insects in the farm and hence can be raised with much ease.

Physical features: There are a few peculiar features in the physical appearance of a turkey. Male has black beard attached to the upper breast regions. This beard is small in females of white turkeys and absent in those of coloured varieties. A fleshy protuberance called snood is present near the base of the beak in both males and females. The head bears caruncles and a large flap of skin below the chin is known as dewlap. Broad Breasted Bronze breed are ordinarily seen in India. Norfolks and the Cambridge are the other important breeds.

Housing: Housing for turkeys may be of range system, confinement or a combination of both. There is a growing trend towards confinement housing. Breeds require a floor space of 0.4 m^2/ bird in confinement and 32.5 m^2/bird in the range rearing.

Feeding: Turkeys have much higher growth rate compared to chickens. Hence they have relatively higher feed and nutrient requirement. Turkey can be fed with all types of dry feed comprising pellets, wet mash, grain and green feed in well balanced combination. They also need plenty of clean water.

Maintenance: Mating ratio in natural flock is 14 hens: 1 tom for small and 12:1 for medium turkeys. The birds should be care fully segregated when they reach the mating stage. Mating takes place during summer. Particular attention should be bestowed in regulating mating. The males should be kept away from their mates in the morning and selectively mated with other birds by turn. Mating in the evening increases the number of fertile eggs. Generally turkey lays eggs on alternate days from six months onwards. In a year, they lay about 150 eggs. The bird attains a weight of 5-8 kg in a year. The birds should not be allowed to fatten too much as it would make them vulnerable for diseases. Changing the breeding stock periodically would be gainful for the farmers.

g. Pigeon rearing

Domestic pigeon *(Columba livia)* belonging to the sub-family Columbia can be reared for its meat. Females are smaller than the males and finer in the head. They hold their tails higher than males and their pelvic bones are spaced wide apart. The males are large, aggressive and make a loud noise. They strut around females during the breeding season. There is difference in the growth rate among the males and females. The male grows faster than the female. Thus, even at an early stage, the gender difference can be easily made based on their body weight.

Feeding: Adults feed the young ones (squabs) with 'pigeon milk', a fluid that is secreted their crop. The pigeon milk looks like thick custard and has high protein content. Pigeons require water for drinking and for bathing to remove external parasites. A pair of breeds drinks about 500 ml of water per day. Pigeons are not fussy eaters, but do best on diets having whole grains. Nutrient requirement of pigeons is crude protein (13.5%), carbohydrates (65%), crude fibre (3.5%). The daily feed intake of an adult pair is about 120 g.

Maintenance: Each pair normally produces 14 squabs in a year. The incubation period of the eggs are 18 days.

Both partners build their nest and take turns to sit on the eggs. The female starts laying eggs again when the squabs are two weeks old. The young ones become independent in 35 days. Squabs of 30-35 days age are best suited for meat purpose since recovery is high with the body weight of 350 g. In addition to the meat, pigeon droppings (25 g per bird per day) can be used as fish feed or enriched manure.

h. Japanese quail

Quail belongs to class Aves, family *Phasianidae* and genus *Cotumir*. Quails, initially experimental birds, have become popular for their eggs and meat. Efforts are being made by Central Avian Research Institute, lzatnagar, to develop the guinea fowl for meat production. Males are lighter than females and are identified by the cinnamon coloured feather on the upper throat and lower breast region. Female is similar to male, in colour except that the feathers on the throat and upper breast are long pointed and much lighter cinnamon in colour.

Maintenance: Quails need small space and are highly suitable for cage rearing. Therefore, it is possible to rear them even in urban areas where space is a major constraint. They are socially and commercially more adaptable due to their small body size, shorter generation interval, highly prolific nature and simplicity in rearing and are more economical. Females lay eggs as early as at the age of 5 weeks. The average age of the first egg production is 40 days. They are in full production by 50 days. They have a capacity to lay 300 or more eggs each per annum. The shell of a typical quail's egg is a mixture of greenish black and brownish speckles or patches of varying sizes. An egg weighs about 10-12 g on an average and is comparatively thin and delicate in nature. Albumen content is 59 to 65%, yolk 28 – 32% and shell 12-14%. The yolk content in relation to albumen is more in a quail's egg than in a chicken's egg. Special care should be given for collecting and handling the egg. The rapid growth rate of the Japanese quail renders it an ideal broiler. At six weeks, quails attain a body weight of 125-150 g with a feed efficiency of 3.0%. Japanese quails are fairly resistant to most of the common diseases affecting chicken. However, they are sensitive to abrupt environmental changes during first and second week of their life.

i. Piggery

Pigs are maintained for the production of pork. They are fed with inedible feeds, forages, certain grain byproducts obtained from mills, meat byproducts, damaged feeds and garbage. Most of these feeds are either not edible or not very palatable to human beings. The pig grows fast and is a prolific breeder, farrowing 10 to 12 pigs at a time. It is capable of producing two litters per year under good management conditions. The carcass return is high at 65-70% of the live weight.

Breeds: Imported breeds of Large White Yorkshire and Landrace are being used widely.

Yorkshire is the most extensively used exotic breed in India. It is a prolific breed having good carcass quality, growth rate and conversion ability. For a small breeding farm or unit, the selection of a herd boar is extremely important. A good boar weighs 90 kg in about 5-6 months and is strong on feet and legs. The mother of the pig to be selected should have large litters of eight pig1lets or more.

Housing: Housing should provide maximum comfort to pigs so that their growth is optimum. There should not be dampness, draft and over heating. Locally available materials can be used for housing. One pig requires about 2.7 sq. m with a wall of 1.2 m height. Eight boars can be kept in 2.7-4.5 sq. m area with 2.4-6.0 sq. m open space.

Feeding: Feed plays an important role in successful pig production. Pigs are the most rapidly growing livestock and suffer more from nutritional deficiencies than the ruminants. Protein, carbohydrates, fats, minerals, vitamins and ample good water form a complete diet for pig. Pigs have a simple stomach; therefore, they must be fed with maximum of concentrates and minimum of roughage. The main ingredients of swine ration are cereals and millets and their byproducts. The fibre content in swine ration should be very low (around 5-6%) for better feed utilization efficiency. Mixed ration should also contain 0.5% of added salt. Swine requires comparatively higher percentage of Ca and P than do cattle or sheep. When pigs are maintained with agricultural/kitchen waste or fish and slaughter house waste, the cost of production remains low. On an average, the consumption of feed is 3.5% of total weight. Feed allowance is calculated as 2.5 - 3.0 kg/100 kg body weight + @ 0.25 kg feed per piglet with the lactating mothers.

Management: As a general rule, well developed gilts weighing about 100 kg, when 12-14 months old, may be used for breeding. The body weight is more important than age at breeding. Sows with low body weight show higher rate of fetal and per-weaning mortality and have been proved as mothers with poor nursing ability. The gestation period is on an average 114 days. Litter size at birth may be 1-16 numbers with the body weight of 1 -25 kg. Normal period between births of piglets is 10-20 minutes. Time taken for the whole process of farrowing ranges from 1.5 to 4.0 hours. Sows are weaned after 40 days. It is advantageous to cull the sow after fifth or sixth litter in a commercial herd. Weaned sows come into heat in 3-10 days after weaning and may be allowed to breed. The boar-sow ratio should be 1: 15. It is profitable to raise two litters from each sow each year.

Mortality in piglets is an important cause of heavy economic loss leading to failure of pig industry. In general, one fourth of piglets farrowed die before they are weaned. Another one tenth is also categorized into stunt or unprofitable group due to disease or parasite infection. Thus, about 60-65% of the piglets farrowed perform as healthy piglets at slaughter age. Death rate is high during farrowing and the first week after farrowing. The farrowing season also determines mortality rate. Mortality in new born piglets is maximum when farrowing takes place in acute cold or hot climate. Therefore, mating should be planned in such a way that farrowing could be avoided in such seasons.

j. Rabbit rearing

Rabbit *(Oryctologus cuniculles)* is reared all over the world in various climatic conditions. Rabbit rearing in India is of recent origin, though hunting of wild rabbits for meat is common. Rabbits can easily be reared with relatively less concentrate feed and with very high reproduction rates. Rabbit meat does not have any religious taboo, unlike pork and meat from cow.

Breeds: Rabbits can be classified into two categories as meat and wool producing ones. Important meat breeds are New Zealand White, Soviet Chinchilla and Kandinrex. Breeds for wool are Angora, Polish Bolamina, Dan and Flirorida.

Housing: Rabbits can be kept in hutches outside or in cages inside. Cages should have door (30 x30 cm) with welded wire mesh at floor and on sides. The area required by rabbit depends on the body weight. For one kg weight, 1.8 sq. m area is needed. Angora rabbits cannot be kept at high temperature and should be restricted to hills. Breeding bucks and adult does are housed single. Castrated male can be kept in group. Young rabbits also live in group till the age of 4 months and thereafter they are separated. Rabbit keeping requires calm and quite areas. Sudden noise can cause death of rabbits. Disturbance by dogs, cats and humans should be kept to the minimum.

Feeding: Rabbit is a non-ruminant herbivorous animal having a simple stomach. Like other animal, rabbit requires protein, carbohydrates, fats, minerals and vitamins as its feed. Rabbits digest the fibre fraction very poorly and a diet having 5-7% crude fibre is adequate. Alfalfa and other forages are well utilized by rabbits. The feed ingredients may constitute maize (20%), alfalfa (25%), wheat bran (40%), groundnut cake (12%), mineral mixture (2%) and salt (1%). The feed mixture should be given twice a day, in the morning and evening. Other grasses and vegetables can be given intermittently in small quantities. Water should also always be provided.

Maintenance: Rabbits of 5-6 months age are ready for breeding and the body weight should not be less than 2.0 kg. The female rabbit is taken to the cage of male for mating. At the time of mating, the male produces a particular noise. After that the female rabbit is taken out of the cage of the male. Early morning is the best time for mating. Rabbits mostly breed from March to July. Gestation period is 28-32 days. The eyes of the young ones (pups) open around two weeks & at 3 weeks, they feed themselves. During this period, the pups should not be often handled. Around four weeks, the young ones are separated from the mother. After two weeks of weaning, the female can be allowed for mating. In a year, it gives birth five times, each time producing about six numbers. Mortality in the kindlings is the most important component that determines the economics of rabbit rearing. In general, mortality in adult rabbit is very low ranging from 3-5%. However, it ranges from 10-15% in kindlings up to the age of 30-42 days. Mating during extreme hot and cold season causes high mortality of kindlings.

k. Bee keeping

Bee keeping is one of the most important agro-based industries which does not require any material from the artisan like other industries. Nectar and pollen from flowers are the materials which are available in plenty in nature. Bee keeping can even be started with a single colony.

Species: There are two species, *Apis cerana indica and A. mellifera* which are complementary to each other but have different adaptations. *A. cerana* is better acclimatized to higher altitudes of the Himalayan region. *A. mellifera* is more profitable in the plains.

Management: The bee- keeper should be familiar with the source of nectar and pollen within his locality. The most important sources of nectar and pollen are maize, sorghum, sunflower and palm. The beginner should start with 2 and not more than 5 colonies. A minimum of 2 colonies is recommended because in the event of some mishap, such as the loss of the queen occurring in one, advantage may be taken with the other. The hives consists of floor- board, brood chamber, super chamber, top cover, inner cover, frame and entrance rod. These parts can easily be separated. The hive may be double walled or single walled. The single walled hive is light and cheap.

The most suitable time for commencing bee keeping in a locality is the arrival of the swarming season. Swarming is a natural tendency of bees to divide their colonies under conditions that are generally favourable for the survival of both parent colony and the swarm. This occurs during the late spring or early summer.

Honey collection: Honey should have good quality to meet the national and international standards. Qualities such as aroma, colour, consistency and floral sources are important. Proper straining and processing are needed to improve the quality of the produce. Honey varies in the proportion of its constituents owing to the differences in the nectar produced by different plants. The nectar collected by bees is processed and placed in comb for ripening. During the ripening, sucrose is converted into glucose and fructose by an enzyme called invertase which is added to it by the bees. Honey is an excellent energy food with an average of about 3500 calories per kg. It is directly absorbed into the blood stream, requiring no digestion.

l. Sericulture

Sericulture is defined as a practice of combining mulberry cultivation, silk worm rearing and silk reeling. Sericulture is a recognized practice in India. The total area under mulberry is 240 lakh ha in the country. It plays an important role in socio-economic development of rural poor in some areas.

The climatic conditions in India are favourable for luxuriant growth of mulberry and rearing silkworms through out the year. The temperature in Karnataka state, major silk producing state in India, ranges from 21.2 to 30°C. Climatic conditions in Kashmir are favourable to rear silk worm from May to October.

Moriculture: Cultivation of mulberry plants is called as 'moriculture'. There are about 20 species of mulberry, of which three are commonly cultivated. They are *Morus alba, M.indica* and *M. latifolia.* The crop can yield well for 12 years, after which they are pulled out and fresh planting is done. Yield of mulberry leaves is 30-40 t/ha/year.

There are four types of silk worm:

Mulberry silk worm - *Bombyx mori*

Eri silk worm - *Philosamia ricini*

Tassar silk worm - *Antheraea mylitta*

Muga silk worm - *Antheraea assami*

Rearing and maintenance: The fertilized moth is covered with an inverted funnel or cellule and eggs are allowed to be laid over a card board. Parasites may be removed by brushing the egg masses with a fine brush. This will also enable to obtain a uniform hatch. In a bamboo tray rice husk is spread. Tender chopped mulberry leaves are added to the tray. The hatched out larvae are transferred to the leaves. It is important to change the leaves every 2-3 hours during the first 2-3 days. The cocoon is constructed with a single reel thread of silk. If the moths are allowed to emerge from the cocoons, the silk thread is cut into pieces. Hence, the pupa are killed 2-3 days before the emergence of moth and processed. The cocoons required for further rearing are kept separately and moths are allowed to emerge from them.

m. Mushroom cultivation

Mushroom is an edible fungi with great diversity in shape, size and colour. Essentially mushroom is a vegetable that is cultivated in protected farms in a highly sanitized atmosphere. Just like other vegetable, mushroom contains 90% moisture with high in quality protein. Mushrooms are fairly good sources of vitamin C and B complex. The protein have 60-70% digestibility and contain all essential amino acids. It is also rich sources of minerals like Ca, P, K, and Cu. They contain less of fat and Carbohydrate and are considered good for diabetic and blood pressure patients.

Varieties

1. Oyster mushroom - *Pleurotus sp.* (var Co 1, Co2)
2. Paddy straw mushroom - *Volvariella volvacea*
3. White button mushroom - *Agaricus bisporus* (var A-II, Horst u3.)

Methods of Production

Oyster mushroom: Take fresh paddy straw and cut into small pieces of 3-5 cm length. Soak them in water for 4-6 hours and then boil for half an hour. Drain the water and dry the straw in shade till it is neither too dry nor wet. Take polythene bags of 60 x 30cm size and make two holes of one cm diameter in the centre of the bag such that they face opposite sides. Tie the bottom of the bag with a thread to make a flat bottom. Fill the bag with paddy straw to 10 cm height. Then inoculate with the spawn. Likewise prepare 4-5 layers of straw and spawn alternatively. The last layer ends up in straw of 10 cm height. Keep this in a spawn running room maintained at a temperature of about 22-28°C and with RH 85-90%. After 15-20 days when the spawn running is completed, cut open the polythene bag and take it to cropping room and allow it to grow for 7 days and harvest the mushroom. Yields about 0.5-1.0 kg/ bed.

Paddy straw mushroom: Cut the straw into long pieces of 60-90 cm and soak in water for 12 hours and sterilize for 15 minutes. Arrange the straw in bundles. Lay the moistened straw bundles on the slightly raised concrete floor or on wooden platform in layers of four bundles width. Spawn or seed the beds simultaneously in each layer either by broadcasting or placing the grain spawn at different spots. Sprinkle grain dhal over each layer on spawn. Don't spawn below the topmost layer. Maintain it at 30-35°C. Harvesting is ready after 25-30 days. Yields about 1-1.5 kg/ bed.

n. Biogas plant

Biogas is a clean, unpolluted and cheap source of energy, which can be obtained by a simple mechanism and little investment. The gas is generated from the cow dung during anaerobic decomposition. Biogas generation is a complex biochemical process. The celluloses material is broken down to methane and carbon-di-oxide by different groups of microorganisms. It can be used for cooking purpose, burning lamps, running pumps etc.

Types of Biogas Plant

Float dome type: Different models are available in this category, *e.g.*. KVIC vertical and horizontal, Pragati Model, Ganesh model.

Fixed dome type: The gas plant is dome shape underground construction. The masonry gas holder is an integral part of the digester called dome. The gas produced in the digester is collected in dome at vertical pressure by displacement of slurry in inlet and outlet. The entire construction is made of bricks and cement. The models available in this category are Janata and Deen-Bandhu.

The selection of a particular type depends on technical, climatological, geographical and economic factors.

Size: The size of the bio-gas plant is decided by the number of family members and the availability of cattle. One cubic metre capacity plant will need two to three animals and 25 kg of dung. The gas produced will meet the requirement of a family of 4-6 members. It is sufficient to have a 2 cubic metre plant to cater to the need of a family of 6-10 members.

Site selection and management: The site should be close to the kitchen or the place of use. It will reduce the cost of gas distribution system. It should also be nearer to the cattle shed to reduce the cost of transportation of cattle dung. Land should be levelled and slightly above the ground level to avoid inflow or run-off of water. Plant should get clear sunshine during most part of the day.

Generation of dung has a direct bearing on the quantity of gas generated. The amount of gas production is considerable higher in summer followed by rainy and winter seasons. Gas production would be maximum at a temperature between 30 to 35°C. If the ambient temperature falls below 10°C, gas production is reduced drastically.

Biogas slurry: Slurry is obtained after the production of biogas. It is enriched manure. Another positive aspect of this manure is that even after weeks of exposure to the atmosphere, the slurry does not attract fleas and worms.

o. Agroforestry

Agroforestry is a collective name for land use systems and technologies in which woody perennial (Trees, shrubs, palms, bamboos, etc.) are deliberately combined on the same land management unit with herbaceous crops or animal either in some form of spatial arrangements or temporal sequence. The objectives of agroforestry are three fold:

i. *Productivity:* Maximizing productivity of both food and wood

ii. *Sustainability:* Conserving and ameliorating the production base. By this, it ensures sustainability and farmer who may not be interested in the conservation goals for their own sake.

iii. *Adoptability:* Notwithstanding the technical elegance and environmental excellence of any agroforestry system, the intended user does not adopt it; it ceases to be a practical proposition.

Benefits of agroforestry

The farmer still uses cow dung as fuel due to shortage of wood. It is estimated that the farmyard manure our produces, equals the production of a thousand fertilizer factories. It is estimated that over 458 million of wet dung is diverted annually which could otherwise fertilize 91 million ha @ 5 t ha^{-1}. Assuming an additional yield of 0.5 t ha^{-1}, an overall increase in yield of 45 m.t can be expected. If the farmer creates fuel resources on his own farm by raising fuel trees the combustion of valuable cow dung could be avoided and the same diverted to the soil.

Improvement and maintenance of soil productivity

Soil is the common d1enominator of life and the serious repercussions of its loss due to water or wind can be understood from the fact that crop yields are depressed as much as 77 percent by removal of just 5.0 to 7.5 cm of surface soil. Removal of one cm of nutrient rich surface soil can result in the export of 24 kg P and 240 kg of N per hectare. Even at the lowest interpretable value in India, it is equal to 2.5, 3.8 and 2.6 million ton of N, P and K, which is greater than annual consumption of these nutrients through fertilizers. Out of the various service functions, the most important is the potential of trees to maintain soil productivity.

Effect of trees on soil properties

Tree has a beneficial effect on soil physicochemical properties, which include improvement of soil structure, increase in nutrient availability and better soil nutrient conservation. Owing to build up of soil fertility below trees, there is invariably a better growth of under storey vegetation compared to areas away from the tree canopy. The nature and amount of fertility improvements vary among species. Thus soil fertility studies under 14 year old tree communities of *Prosopis cineraria, P. juliflora* and bare sites showed an overall improvement in organic matter, total nitrogen and total phosphorous and a slight reduction in pH and EC values under *Prosopis cineraria.*

The increase was confined largely to the upper layer of soil but it was evident that that the build up of soil fertility was maximum beneath *Prosopis cineraria* and *P. juliflora.*

Agroforestry systems

It is a broad term which means raising woody perennials with shrubs and herbs in different combinations in order to utilize the resources in a better way and to improve the production potential of the given land area. It is also called as multiple land use system. The different agroforestry systems are as follows.

i. Agri silviculture

In this system, agricultural crops are intercropped with tree crops in the interspace between the trees. Under this system, agricultural crops can be grown upto two years in protective irrigated conditions and upto 4 years in rainfed conditions. Beyond this period, it is uneconomical to grow cash or grain crops whereas fodder crops, shade loving crops or shallow rooted crops can be grown. Even wider spacing can be adopted without sacrificing tree population for each cultural operation as well as get more sunlight to the intercrop. Performance of tree crops in this system is better compared to monoculture.

ii. Sylvi - pasture

Trees are planted in the poor, degraded soils and fodder grass and legume mixtures are being raised in the interspace of trees. The trees should be multipurpose and provide shade, fodder and fuel and be drought tolerant and leguminous. A typical example in Tamil Nadu is Kangeyam tract, where Acacia leucophloea is being grown and fodder grass Cenchrus or fodder sorghum is raised as intercrop. The tree canopy gives better shade to the cattle and the pods become palatable feed to the cattle and sheep.

iii. Sylvi – Horti – Pastoral System

In this system, trees are grown along with fruit trees and in the inter space, grass and legume mixtures are raised. In Dharwad, Karnataka, encouraging results have been obtained in this agroforestry system. Casuarina was intercropped with fruit tree like Mandarin orange and the performance of orange was excellent compared to other trees like Eucalyptus and *Leucaena sp*. The incidence of pest and disease were negligible on orange when grown with casuarina. By adopting the above agroforestry systems, we can solve the problems of wood, food and fodder and restore the ecological balance. Further, by adopting the above agroforestry system, unemployment problem can be solved to some extent and the socio economic conditions of the farming community can be improved.

3.11 Interaction between components

3.11.1 Rice cum fish culture

This is commonly practiced in coastal areas in the states of Kerala and West Bengal. This system has great scope in West Bengal, Assam, Bihar, U.P, Orissa and A.P where only single long duration rice crop varieties are cultivated.

Central Inland Fishery Research Institute (CIFRI) have indicated that fresh water fishes can be cultivated in paddy fields with ponds constructed within the paddy field. Two crops of tall variety of rice (deep water variety) in *kharif* and HYV (dwarf) in *rabi* are possible and single crop of fish covering both the seasons can be raised.

Types of Trenches suitable for Rice + Fish system

1. Perimeter type
2. Central pond type
3. Lateral trench type

Depth: 1 m - 1.5 m

Suitable composite fish culture (catla, rohu, mirgal and common carp)

Area for fish pond: 1/10th of rice field

Pest control: Pesticides like neem based pesticides, carbonates and selective organophosphates may be used in low doses.

When rice crop is harvested during Nov-Dec, the water recedes from rice field and fishes take shelter in the trenches. During *rabi,* rice field is once again prepared and dwarf varieties are planted.

Average fish yield: 700 – 1000 kg / year/ 25 cent fish pond

Financial return: 20% over monocropping

Rice + Duck + Fish farming

Total ducks : 200 – 320 kg (1 kg weight /duck) / ha pond

Manure : 10 – 15 t / ha / year

Manure collected and applied (or) it may be constructed above the water surface of pond

Production : 3000 – 4500 kg meat / year

Two rice + Fish + Azolla + Calotrophis

Rice varieties: ADT 36, ADT 38 during *kharif* and *rabi* respectively

Azolla microphylla: 2 t / ha (fish feed cum 'N' fixer) applied at 5 DAT and Calotrophis a wasteland weed incorporated @ 12.5 t/ha as GLM at 10 DAT.

NPK fertilizer: 100: 50: 50 kg/ha

Water depth in rice field: 5 cm

Plant protection (Need based): 5 % Neem Seed Kernel Extract

Fingerlings: Catla, rohu, mirgal in equal ratio

Stocking density: 3000/ha

Stocking time: 15 DAT

Feed: Banana pseudo stem with cow dung 1:1 mixture + Rice bran } 3% body weight of fishes

Depth of fish trenches : 1 m depth

Width : 1 m

Area : 10% of rice area to shelter the fishes

Paddy yield : 10,125 kg from two rice crop

Net return : Rs.17488/-

Fish yield : 173 kg

Nitrogen gain : 24 kg/ha

Rice – Fish + Vegetable farming

Vegetable crops like bitter gourd, ribbed gourd are raised along the raised bunds of fish trenches

Yield	Rice yield (kg/ha)	Fish yield (kg/ha)	Vegetable yield (kg/ha)	Net return (Rs./ha)
Rice – Rice + Fish + Vegetables	10094	407	132	19496
Rice – Rice	10285	-	-	15840

Crop + fish (pond system) + poultry + mushroom

i) Crop : Rice – Rice – Maize : 0.90 ha
Rice – Rice - Groundnut

ii) Fish : 0.10 ha
Fish stocking density : 400 Nos.

iii) Poultry bird : 20 Nos. of layers
Poultry dropping : 2 kg / day from 20 birds

iv) Mushroom : 2 kg production capacity/day

Fish + Pig farming

Pig site : Constructed on the embankment of the fish pond

Pig manure : Drained directly to fish pond
No. of pigs : 2-3
Fish yield : 250 - 300 kg

Pig manure contains 80% digestible food for fish.

3.11.2 Crop + livestock (Dairy)

Dairy forms an important component in farming system especially under garden land conditions. Crop + live stock farming system is a broad system involving farming system that are suitable for different ecosystem. Crop + dairy will be an important system for garden land area. While crop + poultry is ideal for wet land system and crop + goat, crop + sheep ideal for dryland system. Various interactions (both competitive and complementary) are discussed here.

Complementary interaction

1. Milch cows produce milk which gives most sustainable income among the income generated from allied enterprises.

2. Cow dung forms a rich resource of organic manure and can be recycled to the field. Approximately 1.2 tonnes of dry cow dung from each animal can be recycled into the field.

3. Maintenance of three milch animals can generate enough cow dung and cow dung treated in anaerobic decomposition, bio gas will be released. The bio gas is a mixture of methane, carbon-di-oxide and traces of nitrogen, hydrogen, sulphide, oxygen and ammonia. Methane constitutes nearly 60% of volume.

4. The calorific value of bio gas is 3600 K Cal/m^3. Biogas can be used for cooking, lighting and as a substitute for diesel. Digested slurry from the biogas plant is a superior organic manure as it has narrow C:N ratio, low per cent of weed seed and colourless.

5. Crop by products are used as feed to cattle. Paddy straw is used to meet the roughage requirement. Pulses and groundnut haulms are also used as cattle feed. In garden land conditions wherever sorghum/maize is raised straw/stover can be used immediately after harvest of crops as green fodder.

Competitive interaction

1. Investment in crop management and dairy management are equally high. The investment in crops is on the cost of tillage, input purchase, labour charges for planting, weeding, harvesting and processing. The investment of dairy involves purchase of feed material like concentrate, labour charges for maintenance and medicine.
2. Capital on fixed cost is also equally high in both enterprises. In crop enterprise digging open/bore well, energisation of pumps, construction of pre-fabricated channel, thrashing floor and in livestock, construction of livestock shed, feeding trough and purchase of livestock are costly.
3. Area for green fodder will pose competition to the crop area and for irrigation water.
4. Similarly in labour scarcity area/period there is competition for labour in both enterprises.

3.11.3 Crop + poultry

Complementary interaction

1. Poultry will reduce the insect population by eating the larvae, destroying egg masses, there by pest damage can be reduced.
2. Poultry dropping if collected properly can be used as organic manure (poultry manure) and it increases the yield of the crops.

3. Reports are available indicating low or sparse weed population in the fields supplied with poultry manure.
4. Poultry dropping becomes useful feeding material for fish grown in fish pond.

Competitive interaction

If bigger units of poultry are maintained there will be competition for capital. Otherwise there is not much competition for capital.

Crop + fish

Aquaculture is terminology used in South East Asian countries for fish farming. Fish culture is done in larger scale in Thailand, Indonesia. Fish culture is common in West Bengal in India.

Complementary interaction

Rice + fish can be raised on the same field either simultaneously or rotationally. In this case decomposing of rice stubble helps the multiplication of plankton for fish, fish fauna enrich the rice field.

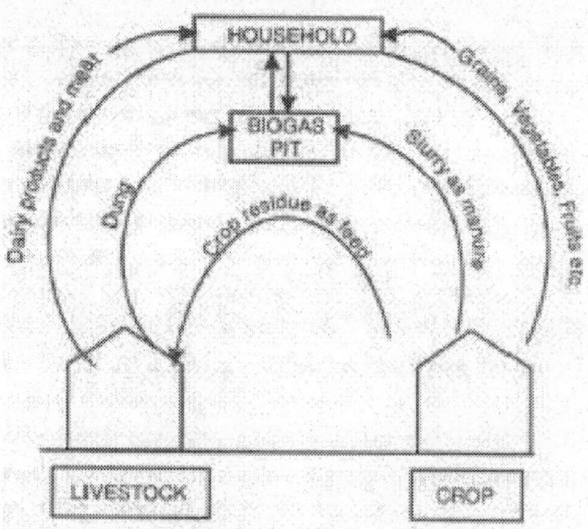

Crop Livestock Farming system

Competitive interaction

There will be intense competition for water by both fishes and rice during summer/dry monsoon period. During this reason alternate crop should be selected.

3.11.4 Crop + goat

Crops and goat association is restricted to dry lands. Hence we must plan the enterprises in such a way that higher rate of complementary interaction is achieved. The competitive interaction should be kept under minimum level. For dry land Tellicherry breed is preferable.

Complementary interaction

- Milk from goat is very nutritious and can be used for human consumption.
- By regular sale of young ones (one or two goats) a regular monthly income of Rs.400-600 can be achieved from one unit goat enterprise (5+1)
- Goat litter can be used as organic manure and can be recycled to the field.

Competitive interaction

Very slight competitive interaction for area is observed between goat and crops enterprises. Goat may require feed @ 1.5-2.0 kg green fodder and 100 g of concentrate for a day per animal. For obtaining green fodder/ dry fodder at least 0.20 ha is required. This much area has to be allocated for raising fodder crops/fodder trees and this area cannot be used for raising any other dry land crops.

Competition occurs for capital-resource of dry land farmers. He has to invest at least 50% of his economic resource for the purchase of Tellicherry goat, construction of shed etc., hence sufficient financial support is needed. Each goat may cost Rs.1000-1200 and put together 6 goats may cost Rs.6000-7200. Cost of construction of shed Rs. 5000. Total cost of goat unit is around Rs.12,000.

3.11.5 Crop + Mushroom

Complementary interaction

- It gives additional income to the farmers
- Daily 1.5-2.0 kg of mushroom can be produced by utilizing 5 kg dry paddy straw/day in a mushroom shed with dimension of 5 x 3 m. This may give an income of Rs. 40-60 per day.

Competitive interaction

Initial cost involved in shed construction.

3.11.6 Crop + Apiary

Complementary interaction

- It gives additional income
- Due to the increased activity of bees, cross pollination of crops and better seed setting are achieved eg. Sunflower

Competitive interaction- Nil

3.11.7 Crop + rabbit

Complementary interaction

Gives additional income to the farmers.

Competitive interaction

Shed cost

3.11.8 Crop + sericulture

Crop + Sericulture will be successful only in upland conditions with irrigation potential. More family labour and skilled labour are needed for successful rearing of silk worm.

Complementary interaction

- Additional income to the farmers
- Dried mulberry leaves can be used as manure. Larval waste and rejects of leaves obtained after silk worm feeding can be used as manure.

Competitive interaction

- Competition for land is high. At least 0.2 ha land is needed for cultivation of mulberry for rearing 100 DFLs / cycle.
- There is competition for water, manures between mulberry and other food crops.

3.12 Evaluation of Farming System Research

Productivity

Production per unit area. To estimate the productivity of a component and compares with the crop component expressed in terms of equivalent crop yield. Further the production estimation itself varies among the interlinked animal component in Integrated Farming System.

Eg. Rice based farming systems

Productivity in-terms of grain yield can be recorded and expressed as kg of rice grain equivalent yield (GEY), calculated as

$$\text{GEY} = \frac{\text{Productivity of component / intercrop(kg)} \times \text{cost of the component / intercrop (Rs/kg)}}{\text{Cost of main crop (Rs/kg)}}$$

Economic analysis

Parameters like cost of cultivation, cost of production, gross and net returns and per day return can be worked out and expressed as Rs. ha^{-1}

Employment generation

- Labour required for various activities in crop production given as man days /ha/year
- A man working for 8 hours in a day is considered as one man day.

- A woman working for the same period is treated as 2/3 man days and computed to man days

Productivity of Livestock components

Fisheries

Fish weight recorded at harvest and expressed as kg / unit area.

Poultry

Egg production per day from birds and expressed as total number per month from the unit.

Pigeon

Growth rate at monthly interval and weight at the time disposal recorded and expressed as kg/unit.

Mushroom

Yield per day and total yield per year from the unit

Water requirement

Water requirement for varying component linkage in the integrated farming systems expressed in ha cm.

Residue addition

The quantity of residue available form each component (kg)

Potential residue additions in terms of N, P and K

Energy efficiency

Energy input and output were worked out for individual components based on the input and output energies and energy efficiency suggested by Mittal *et al.*, (1985) and Gopalan *et al.*, (1976)

Nutritive value

Nutritive values in terms of carbohydrate, protein and fat (kg)

4. INTEGRATED FARMING SYSTEM RESEARCH AND DEVELOPMENT IN TAMIL NADU

In an effort for a holistic integration of different farming enterprises with cropping with the objectives of increasing income and recycling of farm wastes and by-products to sustain the soil productivity under varied agro-ecological situations, studies on integrated farming system involving various components were carried out at different agro-climatic zones of Tamil Nadu since 1985. The approaches was to find out viable components for wetland, irrigated upland and rainfed situations. The identified technologies emanated from the research programmes on integrated farming systems for the last three decades are enormous and comprehensive as detailed below.

4.1 Western zone of Tamil Nadu

The western zone comprises of Coimbatore, Erode, Karur districts, Tiruchengode taluk of Namakkal district, and part of Dindigul and Madurai districts. The total geographical area is 24.89 lakh ha. The zone has undulating topography and the altitude ranges from 160- 2700 m above MSL. The climate ranges from semi arid to sub humid with frequent occurrence of drought. The annual rainfall varies from 524 to 1428 mm with an average of 786 mm. Of this, nearly 49% is received during North East monsoon season and 32% during South West monsoon season. The maximum temperature ranges from 16.2 to 24.5 °C.

The predominant soil types are red non-calcareous, red calcareous, black calcareous, alluvial and forest soils.

In this zone, nearly 66.5% of the lands are irrigated by wells and 26.2% by canals. The major river systems are lower Bhavani Project, Parambikulam Aliyar project, Amaravathi river project and Mettur West Bank Canal system with total command area of 3.11 lakh hectares.

4.1.1 Wetland

Integrated farming systems experiments were conducted at wetlands from 1987-92, 1993-95 and 1998 - 2001 of Tamil Nadu Agricultural University, Coimbatore involving different components *viz.,* poultry, pigeon, goat, fishery and mushroom. Coimbatore receives the normal total rainfall of 678.2 mm in 46.4 rainy days. The soil of the experimental is moderately drained clay loam. Water requirement of the components in the farming systems and irrigation water requirements of the experimental plots was met from the borewell situated near the experimental field.

Cropping + poultry + fish culture + mushroom

Integrated farming systems involving poultry, fish culture and mushroom cultivation were taken from 1987-1992. In wetland, the normal cropping programme followed is Rice - Rice - Green manure / Pulse. But in the integrated approach, a modification was made in the existing cropping pattern by including crops like maize, groundnut and gingelly. Cropping was undertaken in 0.36 ha and 0.04 ha was allotted for fish pond and the poultry shed was placed above the pond. The polyculture fingerlings are preferred for this system as the growth rate of fish within a year is found more economical than monoculture. The fingerlings were released in the pond as per the recommended stocking density of 10000 numbers per hectare of ponded water. Accordingly, 400 fingerlingss were released in the pond of 0.04 hectare area. No separate feed was applied to the pond; instead, the poultry droppings from the poultry shed placed well above the farm pond formed the feeding material. Twenty Bapkok chicks were used in the proposed programme to meet the feed requirement of fingerlings released in the 10 cent pond.

In this integrated system, the waste products of rice formed 33 % of the poultry feed and the feed cost could be reduced by 57.5 per cent by substituting rice grain, maize, oil cakes of crop component and the lucerne grown around the fish pond. In commercial poultry farm, cost of production per egg will be around 42 paise. By way of reducing the feed cost, the cost of production per egg under the integrated farming system was reduced to 19 paise only. The poultry unit of 20 chicks yielded 4531 eggs with 29 kg of culled birds valued Rs. 2520. The gross income from the integrated farming system was Rs.70619 ha^{-1} $year^{-1}$ and that of the conventional cropping system was Rs.33446 ha^{-1} $year^{-1}$ (Table 2 and Model 1 & 2). Of the income from integrated farming system 59.3 per cent was from cropping, 8.7 per cent from poultry, 7.4 per cent from fish culture and 24.6 per cent from mushroom. The additional net income realized from the integrated farming system was Rs.18360 ha^{-1} $year^{-1}$ (Rangasamy, 1995).

Table 2. Economics and employment generation in Integrated Farming System

System	Gross income (Rs./ha)	Expenditure (Rs./ha)	Net income (Rs./ha)	Employment (man days/ ha./year)
Crop + Poultry + fish + mushroom	70619	34673	35946	1383
Crop cultivation alone	33446	15860	17586	930
Additional benefit	37173	18813	18360	453

(Rangasamy, 1995)

Crop + poultry/ pigeon + fish + mushroom

In this system, the component of integrated farming system involved were crop + fish + mushroom, crop + poultry + fish + mushroom and crop + pigeon + fish + mushroom and was taken from 1993 - 1995. The efficiency of the component linkages was evaluated predominantly on the basis

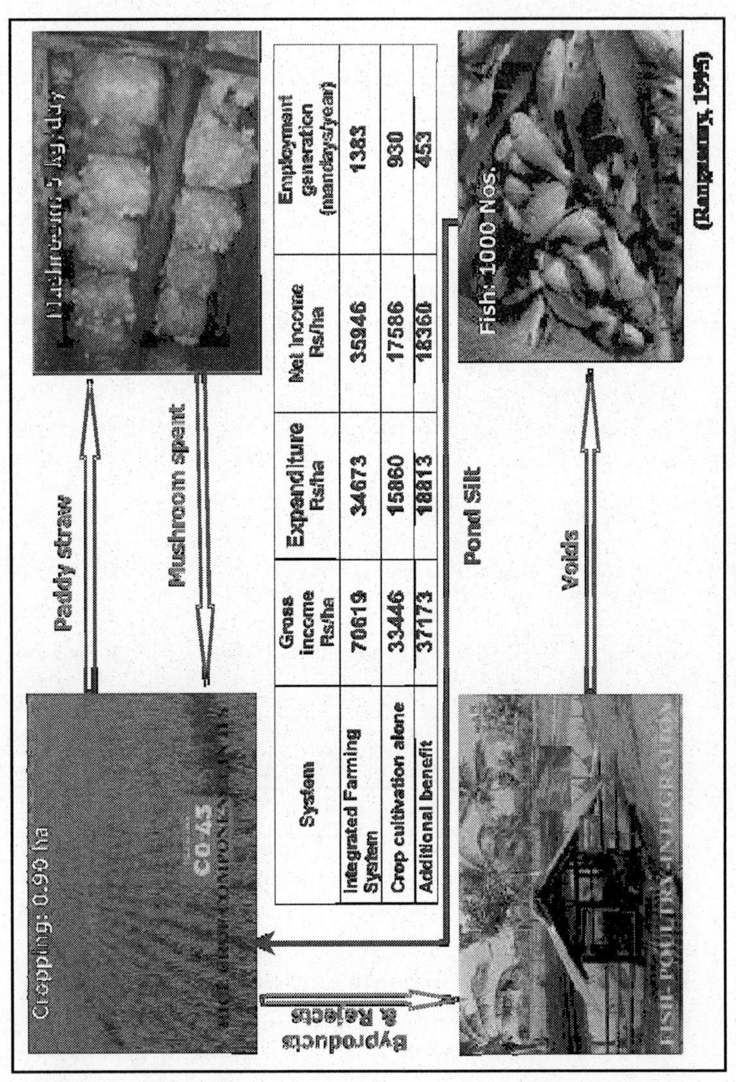

System	Gross income Rs/ha	Expenditure Rs/ha	Net Income Rs/ha	Employment generation (mandays/year)
Integrated Farming System	70619	34673	35946	1383
Crop cultivation alone	33446	15860	17586	930
Additional benefit	37173	18813	18360	453

Model 1. Integrated Farming System for Wetland

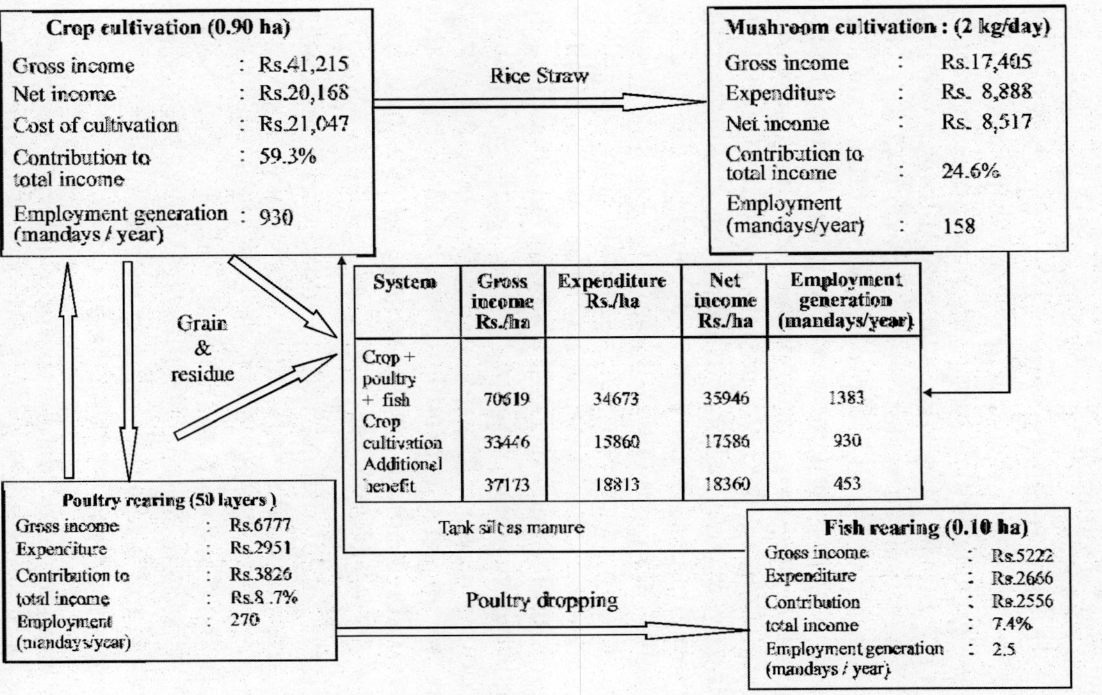

Model 2. Resource Recycling with Productivity and Profitability Linkages in Lowland Integrated Farming Systems (five years mean)

of productivity, its income and employment generation with the possibility of utilizing recycled organic wastes as nutrient to enrich the soil fertility.

Experimental results on enterprise linkage for low land farming systems revealed that rice - soybean - sunflower and rice - gingelly - maize cropping systems each in 0.18 ha with pigeon (40 pairs), fish (400 polyculture fingerlings in 0.04 ha of ponded water) and mushroom (2 kg/day) was best in obtaining higher productivity in the system as a whole than the conventional cropping system being popular with rice-green gram-maize (0.20 ha) and rice - sunnhemp - maize (0.20 ha) cropping systems (Model 3). Cropping + pigeon + fish + mushroom integration earned the highest gross and net returns with better per day returns and benefit cost ratio (Plate 1). Integration of cropping with pigeon + fish + mushroom generated the highest employment of 798 man days ha^{-1} $year^{-1}$ with 1.16 man days from the allied enterprises linked with equi-distribution throughout the year. Integration of poultry + fish + mushroom + cropping applied with recycled poultry manure sustained the productivity of soil through the additions of bio resource residue with better NPK nutrients supply potential (Model 4).

The growth of fish cultures included was comparable among artificial feeding and the treatments with poultry and pigeon droppings feeding. Fish production was higher with artificial feeding than with other two methods (Model 5). However, the net returns and benefit cost ratio were higher in the fish fed with poultry dropping followed by pigeon dropping (Jayanthi, 1995).

To enhance and sustain the productivity, economic returns, employment generation for the family labour round the year and soil fertility with environmental protection, integration of rice-gingelly-maize and rice- soybean - sunflower cropping each in 0.45 ha with recycled poultry

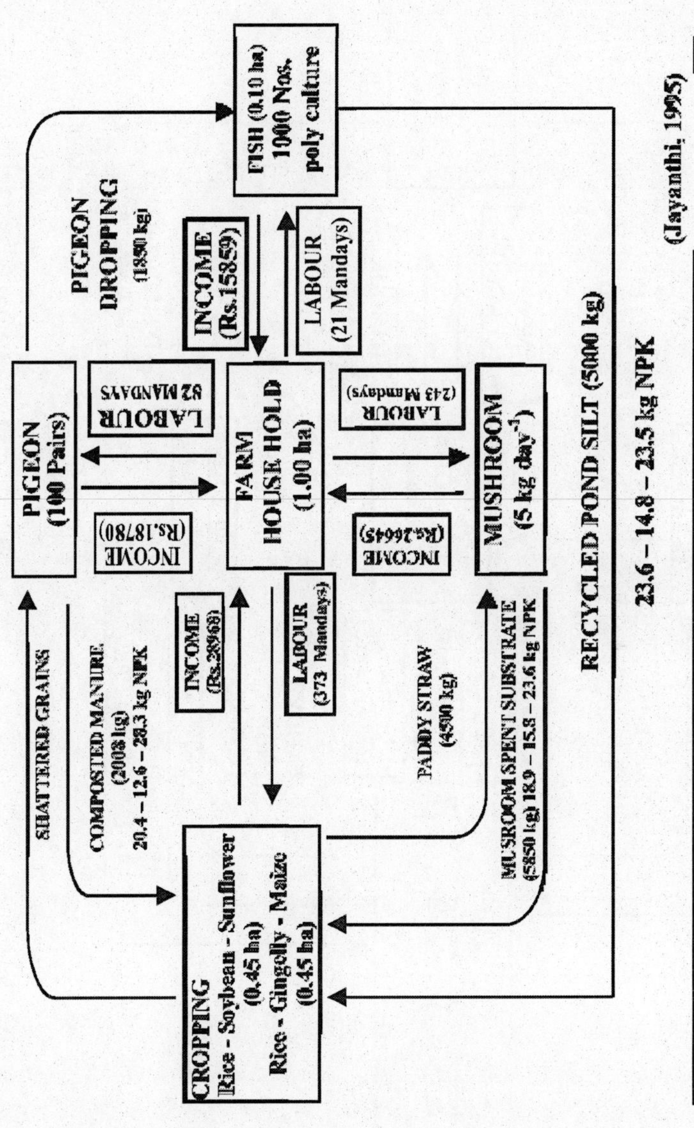

Model 3. Resource Recycling in Crop + Pigeon + Fish + Mushroom Farming System

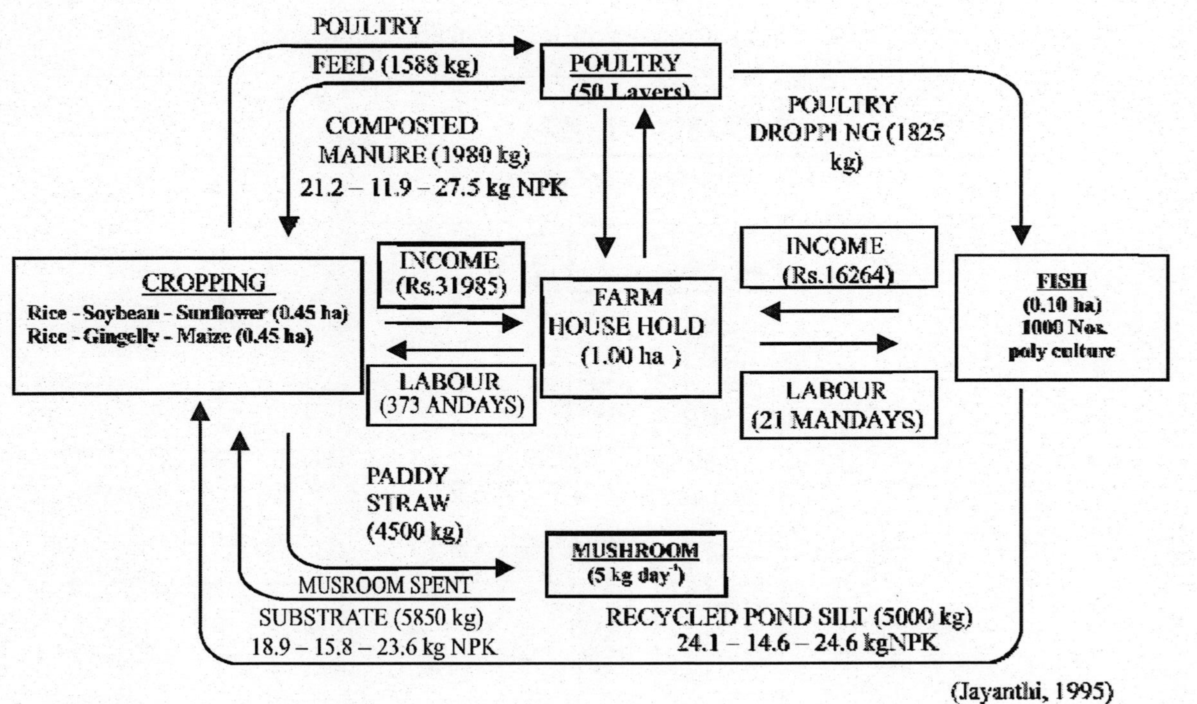

(Jayanthi, 1995)

Model 4. Resource Recycling in Crop + Poultry + Fish + Mushroom Farming System

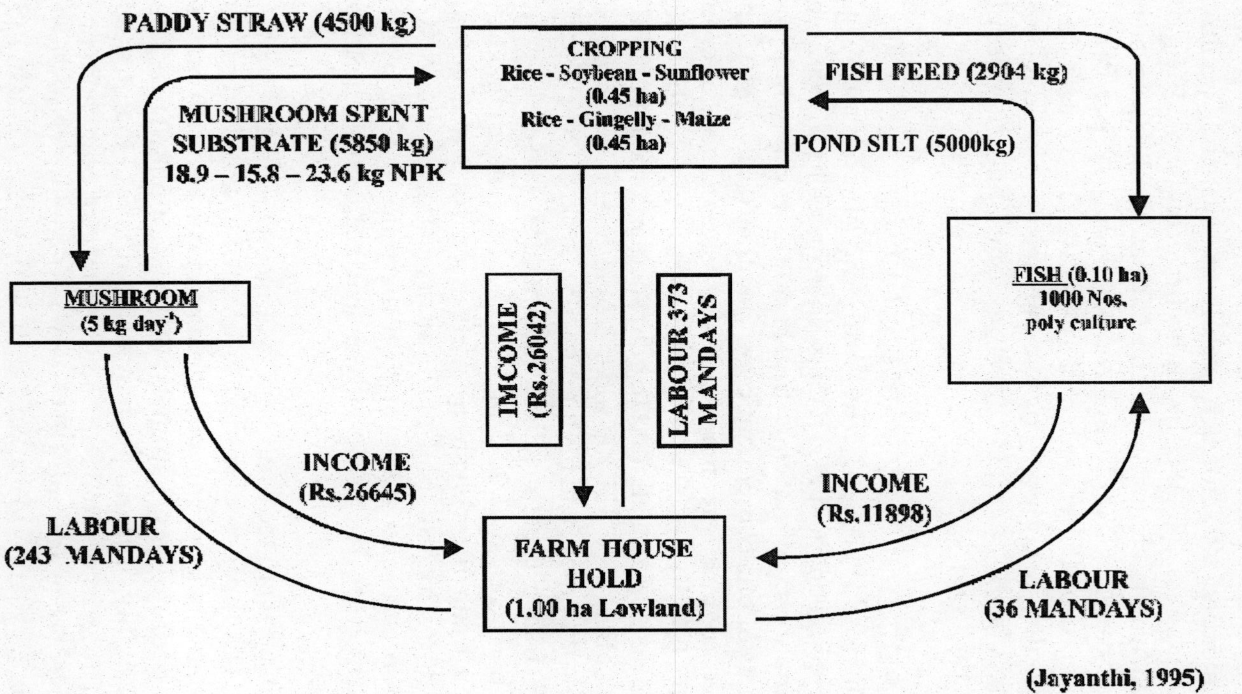

Model 5. Resource Recycling in Crop + Fish + Mushroom Farming System

manure as fish pond silt to rice and 75 per cent of the recommended NPK to each crop in the system + poultry (50 layers) + fish (1000 polyculture fingerlings in 0.10 ha of ponded water) comprising catla (20 per cent), silver carp (20 per cent), rohu (20 per cent), mrigal (15 per cent), common carp (15 per cent) and grass carp (10 per cent) fed with poultry dropping + oyster mushroom (5kg/day) for the lowland farmers having one hectare farm (Jayanthi,1995).

Cropping + poultry / pigeon / goat + fishery

During 1998-2001, the study involved cropping, poultry, pigeon, goat and fishery enterprises in all possible combinations, with a view to recycle the residue and by-products of one component over the other (Plate 2 &3). In one hectare farm, an area of 0.75 ha was assigned for crop activity, 0.10 ha for growing fodder grass to feed the goat unit (20+1), 0.03 ha allotted to goat shed and the remaining 0.12 ha allotted to 3 fish ponds. Three integrated farming systems *viz.*, crop + fish + poultry (20 Bapkok layer birds), crop + fish + pigeon (40 pairs) and crop + fish + goat (Tellicherry breed of 20 female and 1 male maintained in 0.03 ha deep litter system) were tried for three years. Polyculture fingerlings of 400 numbers (catla, rohu, mirgal/ common carp and grass carp) in the ratio of 40:20:30:10, respectively, reared in 3 ponds of size 0.04 ha (depth of 1.5 m) each.

Fishes were fed with poultry, pigeon (700 kg poultry/ pigeon droppings obtained from 20 Bapkok layers/40 productive pairs of pigeon) sheltered over two fish ponds and goat droppings (3 animals - 800 kg droppings) to assess the feasibility of rearing fish by using different manures as feed. Under Integrated Farming System, cropping sequence includes (i) sugarcane (Planted) - sugarcane (Ratoon) - banana (3 years) (ii) banana - turmeric - rice - banana (3 years) and (iii) maize - rice - sesame - sunnhemp (annual) each in 0.25 ha and bajra - napier grass + desmanthus (perennial) in 0.10ha.

Conventional cropping system comprising (i) rice -rice - black gram (ii) maize - rice - black gram (iii) maize - rice - sunnhemp and (iv) rice - rice - sunnhemp each in 0.25 ha, as practised by the farmers was taken up for comparison .

To sustain the productivity of soil through integrated nutrient supply, recycled poultry, pigeon and goat manures and composted crop residue (banana waste and sugarcane trash) as vermicompost each @ 6.25 t/ha were tested along with 100,80 and 60% of the recommended fertilizer for the sequences of cropping.

Integrated farming system provides an opportunity to increase economic yield per unit area per unit time by virtue of intensification of crop and allied enterprises. Research results on integrated farming system for three years revealed that integration of crop with fish, poultry, pigeon and goat resulted in higher productivity than cropping alone under lowland. Crop + fish + goat integration recorded higher rice grain equivalent yield of 39610 kg/ha (mean over three years) than other systems (Table 3). Similarly, as an individual animal component, the goat unit (20 + 1) gave the highest productivity of 8818 kg (mean over three years). This could also provide 11.0 t of valuable manure apart from supplementing the feed requirement of 400 numbers of fish. While assessing the feasibility of rearing fish by using poultry, pigeon and goat droppings as feed, the fish fed with poultry droppings resulted in higher fish yield (825 kg / 0.04 ha ponded water) than the other two sources of feed (Plate 4).

The highest net return of Rs.131118 and per day return of Rs.511 ha^{-1} were obtained by integrating goat + fish + cropping applied with recycled fish pond silt enriched with goat droppings (Table 4). Higher net return of Rs.3.36 for every rupee invested was obtained by integration of pigeon

Table 3. Productivity of integrated farming system and its linked components

Farming systems	*Component Productivity (kg)					System productivity (kg/ha)	Per cent increased over CCS
	Crop	Poultry	Pigeon	Fish	Goat		
FS_1 -Cropping alone	12995	-	-	-	-	12995	-
FS_2 - Crop + Fish + Poultry	26352 (89.0)	1205 (4.1)	-	2052 (6.9)	-	29609	128
FS_3 - Crop + Fish + Pigeon	24854 (85.2)	-	2545 (8.7)	1774 (6.1)	-	29173	124
FS_4 - Crop + Fish + Goat	25725 (68.3)	-	-	1975 (5.2)	9979 (26.5)	37679	190

(Jayanthi, 2002)

* Productivity is expressed in rice grain equivalents yields. Figures in the parenthesis indicate per cent contribution of each component

+ fish + cropping applied with recycled fish pond silt enriched with pigeon droppings.

The employment opportunity was also increased to 576 man days ha^{-1} year^{-1} by integrating fish + goat in the cropping as against cropping alone (369 man days ha^{-1} year^{-1}) (Table 5). Combining cropping with other allied enterprises would increase labour requirement and thus provide scope to employ family labour round the year.

Integration of crop with fish and poultry resulted in higher fish productivity under lowlands. The poultry, pigeon and goat droppings were utilized as feed initially and at the end of a year after the fish harvest, about 4500 kg of settled silt from each pond were collected. The pond silt was utilized as organic sources to supply sufficient quantity of nutrients to the crops. The nutrient contents of raw animal manures and settled silt collected from different fish ponds are furnished in Table 6 and Models 6, 7 and 8.

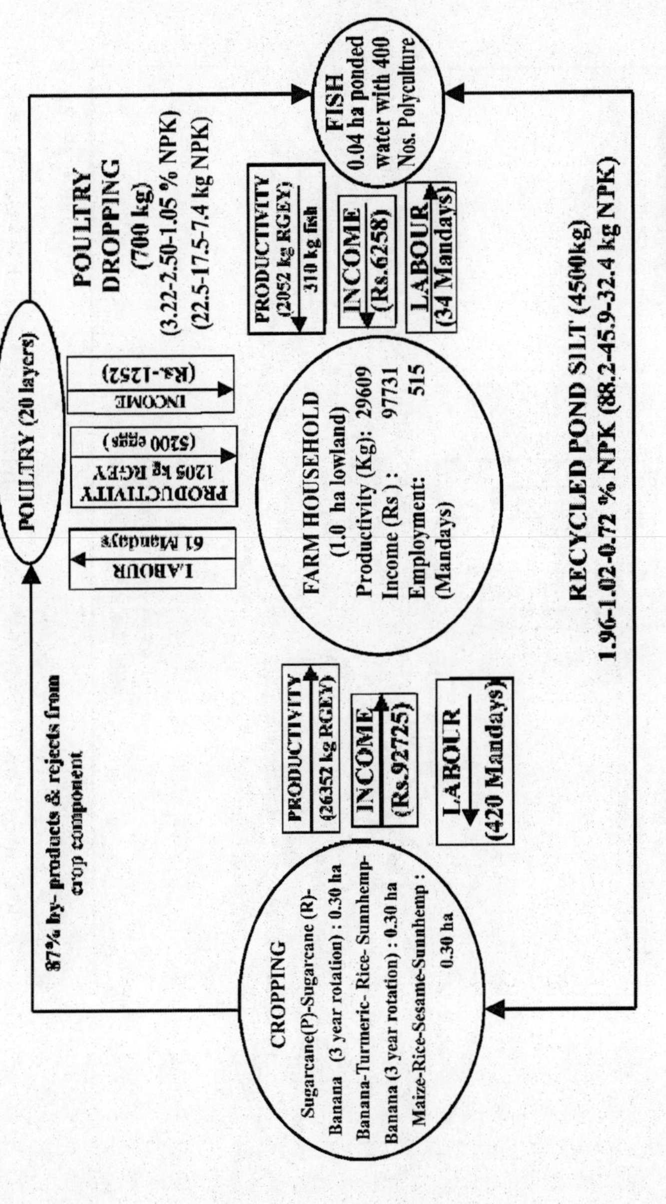

Model 6. Resource Flow in Crop + Poultry + Fish in Integrated Farming System

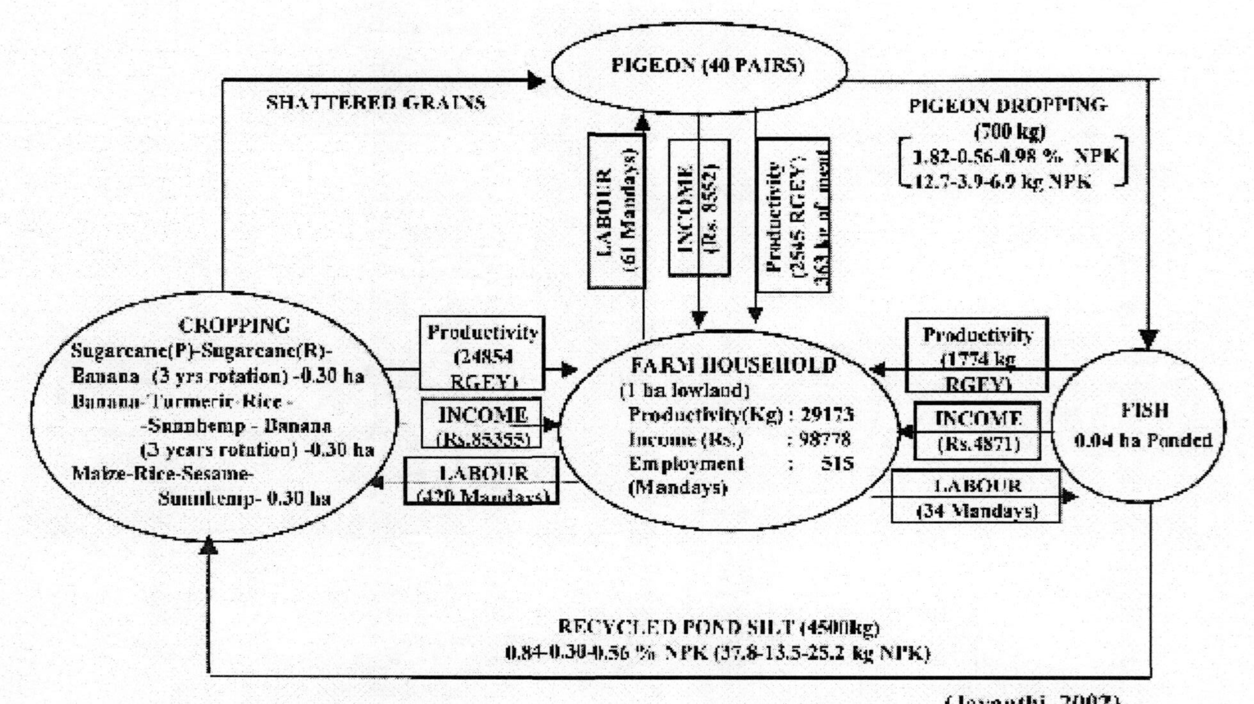

Model 7. Resource Flow in Crop + Pigeon + Fish Integrated Farming System

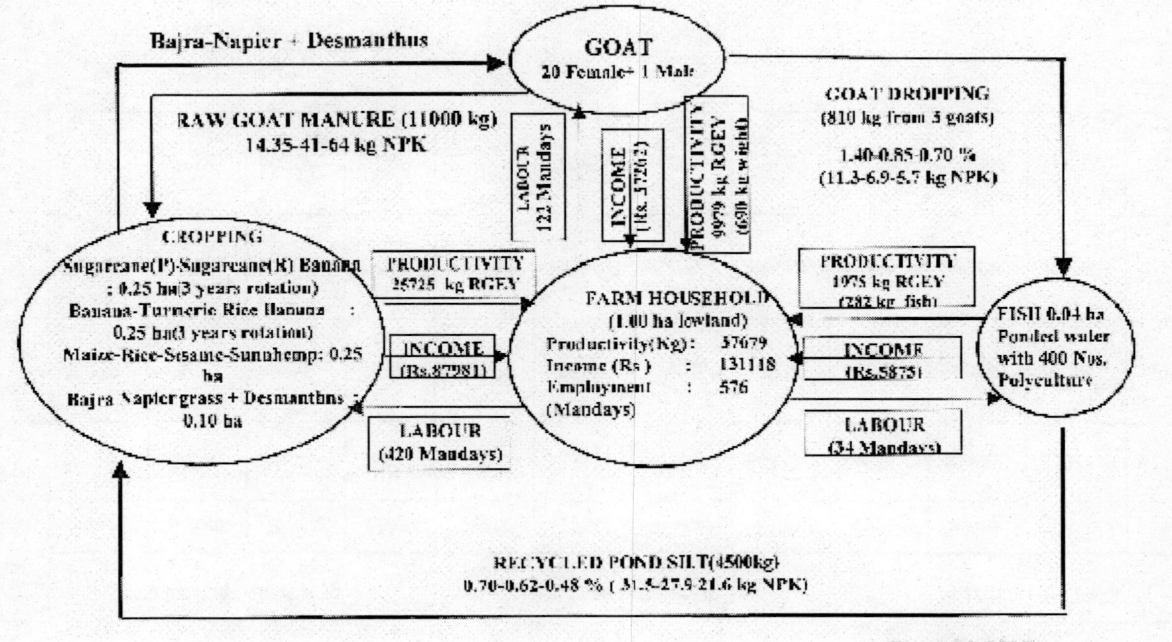

Model 8. Resource Flow in Crop + Goat + Fish Integrated Farming System

Table 4. Economic analysis of Integrated Farming Systems

Farming Systems	Production cost (Rs ha^{-1})				Gross return (Rs ha^{-1})				Net return (Rs ha^{-1})			
	I year	II Year	III year	Mean	I year	II Year	III year	Mean	I year	II Year	III year	Mean
Cropping alone	16905	32939	33623	27822	48826	73399	72700	64975	31921	40460	39077	37153
Crop + Fish + Poultry	60027	42628	42255	48303	183093	135497	119515	146035	123066	92869	77260	97731
Crop + Fish + Pigeon	58842	41177	41250	47090	193358	130191	114055	145868	134516	89013	72805	98778
Crop + Fish + Goat	63677	52689	50282	55549	211566	179351	169085	186667	147889	126662	118804	131118

Table 5. Employment generation in Integrated Farming Systems

Farming systems	Employment generation (mandays)					
	Crop	Poultry	Pigeon	Fish	Goat	Total
1998-1999						
Cropping alone	369	-	-	-	-	369
Crop + Fish + Poultry	415	61	-	34	-	510
Crop + Fish +Pigeon	415	-	61	34	-	510
Crop + Fish + Goat	415	-	-	34	122	571
1999- 2000						
Cropping alone	369	61	-	-	-	369
Crop + Fish + Poultry	424	-	-	34	-	519
Crop + Fish +Pigeon	424	-	61	34	-	519
Crop + Fish + Goat	424	-	-	34	122	580
2000-2001						
Cropping alone	369	-	-	-	-	369
Crop + Fish +Poultry	420	61	61	34	-	515
Crop + Fish + Pigeon	420	-	-	34	-	515
Crop + Fish +Goat	420	-	-	34	122	576

Mean over three years (1998-2001)

Particulars	1998-1999	1999-2000	2000-2001	Mean
Cropping alone	369	369	369	369
Crop + Fish +Poultry	510	519	515	515
Crop + Fish + Pigeon	510	519	515	515
Crop + Fish +Goat	571	580	576	576

(Jayanthi, 2002)

Twenty fowls in the poultry unit and 40 productive pairs in the pigeon unit voided 700 kg of droppings with the nutrient potential of 22.5, 17.5 and 7.4 kg and 12.7, 3.9 and 6.9 kg of N, P_2O_5 and K_2O, respectively, but when recycled through fish pond, nutrient contents were enhanced by three folds. Similarly, three goats produced 810 kg of voiding contributing 11.3, 6.9 and 5.7 kg of N, P_2O_5 and K_2O nutrients, which were further enhanced by three folds through recycling. The additional nutrients gained by recycling were the highest with poultry manure with 65.7, 28.4 and 25.0 kg N, P_2O_5, and K_2O, respectively than with goat/pigeon manure (Jayanthi, 2001).

The system as a whole provided an opportunity to make use of produce/waste materials of one component as input on another at the least cost/no cost at farm level. To enhance the productivity, sustain the income and employment generation for family labour, integration of crop with fish + goat/pigeon/poultry could be adopted than cultivating the crop alone under lowland farms.

Table 6. Nutrient value of recycled poultry/ pigeon/ goat manure

Particulars	Poultry	Pigeon	Goat
Birds /animals used to satisfy the feed requirement of 400 fingerlings	20 layers	40 pairs	3 animals
Quantum of dropping received in an year	700 kg	700 kg	810 kg
Silt cleared after one year from 0.04 ha pond	4.5 t	4.5 t	4.5 t

Nutrient	Raw poultry dropping		Pond manure		Additional nutrient gained (kg)
	%	kg/ 700kg	%	kg/ 4500 kg	
N	3.22	22.5	1.96	88.2	65.7
P_2O_5	2.50	17.5	1.02	45.9	28.4
K_2O	1.05	7.4	0.72	32.4	25.0

Nutrient	Raw Pigeon dropping		Pond manure		Additional nutrient gained (kg)
	%	kg/ 700kg	%	kg/ 4500 kg	
N	1.82	12.7	0.84	37.8	25.1
P_2O_5	0.56	3.9	0.30	13.5	9.6
K_2O	0.98	6.9	0.56	25.2	18.3

Nutrient	Raw Goat dropping		Pond manure		Additional nutrient gained (kg)
	%	kg/ 810kg	%	kg/ 4500 kg	
N	1.40	11.3	0.70	31.5	20.2
P_2O_5	0.85	6.9	0.62	27.9	21.0
K_2O	0.70	5.7	0.48	21.6	15.9

(Jayanthi, 2001)

Rice + Fish + Azolla + Calotropis Farming

A field study was conducted at low-lying wetlands of Lower Bhavani Project Area, Agricultural Research Station, Bhavanisagar from 1992 - 1994 with a view to evaluate the feasibility and economics of rice-fish-azolla farming with calotropis as green leaf manure. Experiment was conducted with two rice crop seasons (kharif and rabi) planted with rice varieties ADT 36 and ADT 38 with a total duration of eight months with one fish crop. *Azolla microphylla* was applied at the rate of 2.0 t/ha as fish feed cum N fixer, five days after transplanting. Calotropis was incorporated (12.5 t/ha) as green leaf manure, 10 days before transplanting. Fertilizer was applied at the recommended level (100:50:50 kg NPK/ha). Irrigation was scheduled to maintain 5 cm depth of water throughout the crop growth period. Need based plant protection measures were taken with 5% neem seed kernel extract.

Fingerlings of catla, roghu and mrigal were stocked in equal ratio with a total of 3000 numbers/ha, 15 days after transplanting. Banana pseudostems with cow dung (1:1) mixture and rice bran were fed at the rate of 5% of the fish body weight. Field trenches were provided with 1.0 m depth and 1.0 m width occupying 10% of the rice area to shelter the fish.

The data presented in Table 7 revealed that in lowlying wetlands, the most profitable farming system was rice - rice - azolla - calotropis + fish farming. There was slight decline in rice yield under rice-rice-fish farming due to the

Table 7. Yield and economics of different farming system (Two years mean)

System	Paddy grain yield kg/ha	Fish yield kg/ha	Net return Rs./ha	Additional income Rs./ha
Rice- Rice	10099	-	15299	-
Rice-Rice-azolla+fish	9870	163	16531	1232
Rice-Rice-azolla Calotropis+fish	10125	173	17488	2189

(Balusamy and Shanmugasundaram, 1994)

presence of fish trenches occupying 10% of the area. However, the yield reduction can be made to normal or even higher when azolla and calotropis were introduced.

The soil fertility status indicated that in general there was an increase in the available nutrient after completion of cycle (Table 8). Even after heavy depletion, through crop removal, the nutrients status could be replenished by the addition of organic residues. This might be due to degradation of azolla, fish manure and fish feed under these systems (Balusamy and Shanmugasundaram, 1994).

Table 8. Nitrogen budget in rice-fish-azolla farming system

System	Initial N (kg/ha)	N added (kg//ha)	Crop removal (kg/ha)	Available (N kg/ha)	Net gain or loss kg/ha
Rice-Rice	227	100	82	219	- 8
Rice-Rice-azolla+fish	227	129	89	245	+18
Rice-Rice-azolla+Calotropis+fish	221	190	93	251	+24

(Balusamy and Shanmugasundaram, 1994)

Rice + Fish + Vegetable Farming

For effective restoration and recycling of organic wastes, an experiment was taken to integrate rice-vegetable and fish culture in wetlands at Agricultural Research Station, Bhavanisagar. The objective is to maximize the return per unit area and also efficient recycling of crop residue along with fish culture.

The experiment consisted of rice alone, rice - rice + fish, rice - rice -vegetable + fish farming systems. The experiment was continued for two seasons with a total duration of eight months. The study area was 0.25 ha/system. Field trenches (1 m deep x 1 m wide) were provided for fish shelter. Catla, rohu and mrigal were stocked 15 days after transplanting at an equal ratio of 3000 fingerlings/ha. Vegetables crops *viz.,* bitter gourd, ribbed gourd, beans, etc., were raised along the raised bund of fish trenches.

The results (Table 9) indicated that the rice - rice - vegetable + fish culture was more remunerative and it was possible to get two rice crop, one vegetable crop and one fish crop within eight months in rice field under two season paddy area (Balusamy and Shanmugasundaram, 1994).

Table 9. Productivity and economic contribution of components in rice-rice-vegetable + fish culture farming systems

System	Rice yield kg/ha		Fish yield kg/ha		Vegetable yield kg/ha		Net income Rs./ha	
	I	II	I	II	I	II	I	II
Rice – Rice	10108	10285	-	-	-	-	15328	15840
Rice - Rice+ Fish	9834	10040	-	-	105	124	16224	17504
Rice- Rice + Fish +Vegetable	9904	10094	450	407	108	132	17888	19496

I : First year (1992- 93) II : Second year (1993 - 94)

(Balusamy and Shanmugasundaram, 1994)

Crop + fish culture (Pond system)

By integrating the fish culture with rice, increased returns could be obtained. Further, the nutrients leached out from rice fields could be effectively utilized for fish culture. With a view to study the feasibility of rice cum fish culture under lowlying wetlands, an experiment was conducted at Agricultural Research Station, Bhavanisagar.

In one acre holding conventional cropping system (rice - rice - black gram) practised by the farmers was followed. In another one acre plot, integrated farming system (rice - rice - maize + fish culture) was practised in an area of 0.85 acre with crop husbandry and 0.15 acre with fish culture. Fish pond size of 600 m^2 was dug to a depth of 2.0 metre. The fish fingerlings of catla, mrigal and common carp were stocked in the fish pond to a density of 3000 nos / ha. The excess water in the paddy fields was drained once in four days from 10 days after transplanting of rice. After draining, irrigation was given upto 5 cm depth each time. Whenever fertilizer was applied by top dressing, the subsequent drainage was done six days after fertilizer application. The same practice was followed for plant protection chemicals application. Supplementary feeding was given to the fish @ 5% of body weight of fishes.

The comparative economics showed that rice - rice - maize + fish culture recorded maximum net returns and it increased over by Rs. 6383 and Rs. 5795 /ha over conventional cropping of rice-rice-black gram during first and second years respectively (Table 10). Drained water increased the yield of fishes and this might be due to increased nutrients available in the drained water pond as revealed by increased plankton population (Balusamy and Shanmugasundaram 1994).

Rice - Fish + Azolla Farming System

Field experiment was conducted at ARS, Bhavanisagar to develop an integrated N management practices for rice -

Table 10. Economics of fish culture in drained water pond of rice based cropping

Systems	Net return (Rs/ ha)		Additional Income (Rs/ ha)	
	1991-1992	1992-1993	1991-1992	1992-1993
Rice - Rice - Maize + Fish	36716	28270	6383	5795
Rice - Rice – Blackgram	29793	22485	-	-

(Balusamy and Shanmugasundaram, 1994)

fish - azolla farming in wetland. Farming systems consisted of rice - rice + fish and rice - rice + azolla + fish and two levels of N (100 and 75 per cent recommended) with and without green leaf manure (*Sebania rostrata*) applications. In the rice - fish system, rice and fish crops were raised together (synchronous system) in rice field. Field trenches were provided with 1.0 m depth and 1.5 m width occupying 10 per cent of the rice area, for sheltering the fish. *Azolla microphylla* was grown in rice field throughout the cropping period (Plate 5). The dual culture method of growing azolla with rice has gained widespread adoptability because standing water is available in rice field from seedling to panicle maturity in lowland rice fields and is effectively used as biofertilizer for rice. Azolla cultivation in rice field can improve the fish food. Fish culture in rice fields loosens the soil as a result of their free movement in water body and thus aerating the soil, enhances the decomposition of organic matter and promotes release of nutrients from soil. The excreta of fish directly fertilize the water in rice fields leading to increase in utilizable source of N to the rice crop. Integration of allied components like azolla + fish with rice in lowland farming could provide wider scope for bio resources recycling (Fig. 6).

Rice - rice - azolla + fish farming with 75 per cent recommended N as well as incorporation of green leaf

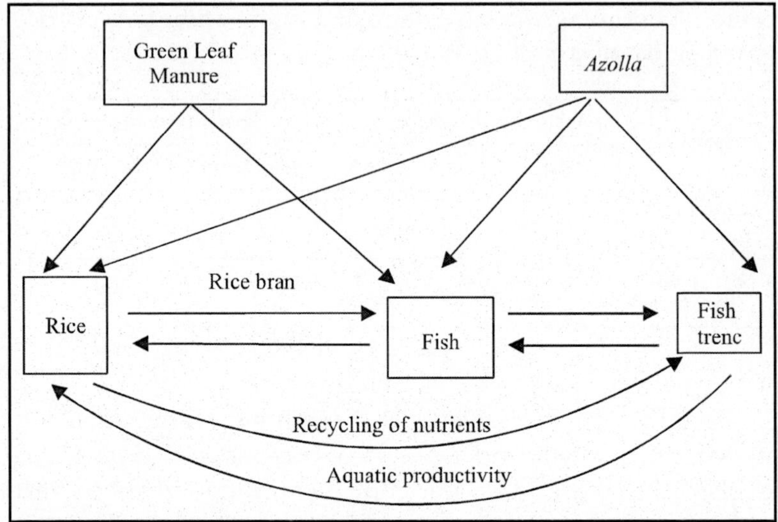

Fig. 6. Bio resource flow in Rice Fish –Azolla integrated farming system

manure resulted in higher productivity with increased economic returns and improved the soil fertility through recycling of organic residues. The quantum of organic residue addition and N added through recycling were higher in rice - rice - azolla + fish farming with *Sesbania rostrata* incorporation (Table 11 and12). The unutilized fish feed, decayed azolla and fish excreta settled at the fish trench bottom had a higher nutrient value, which can be recycled to enrich the soil (Balusamy, 1996).

4.1.2 Upland with supplemental Irrigation

Crop + Dairy +Biogas + Silviculture

Integrated farming system experiments were conducted at irrigated uplands of Tamil Nadu Agricultural University, Coimbatore from 1987-92 and 1999 - 2001. The soil type of the experimental field was clay loam with low in available N, medium in available P and high in available K. The major source of water was met from deep bore well. The major cropping system followed under irrigated

Table 11. Azolla biomass (t ha^{-1}) and N addition (kg ha^{-1}) in Integrated Farming Systems.

Treatments	Biomass added				N addition			
	1992-93		1993-94		1992-93		1993-94	
	I Crop	II Crop	I Crop	II Crop	I Crop	II Crop	I Crop	II Crop
Rice-Rice+Azolla + Fish + 100%N	27.6	17.1	19.1	32.5	54.6	33.9	37.8	64.4
Rice-Rice+ Azolla + Fish + Green leaf manure (GLM) + 100%N	28.2	17.9	20.7	33.6	55.8	35.4	41.0	66.5
Rice- Rice+Azolla + Fish + 75%N	28.0	17.5	20.2	32.7	55.4	34.7	40.0	64.7
Rice-Rice+ Azolla + Fish + GLM +75%N	27.9	18.0	20.5	33.2	55.2	35.6	40.6	65.7

Table 12. N added through recycling of organic residue (kg ha^{-1})

Treatments	N applied through residue recycling					Total
	Fish trench	I crop	II crop	Azolla	GLM	
Rice-Rice	0	9.2	9.9	0	0	19.1
Rice-Rice+Fish +100%N	12.9	8.7	9.6	0	0	31.2
Rice-Rice+Azolla+Fish +100%N	16.9	9.0	10.5	102.2	0	138.6
Rice-Rice+Fish+GLM+100%N	13.4	9.2	11.1	0	19.4	53.1
Rice-Rice+ Azolla + Fish + GLM + 100%N	17.3	9.2	11.3	107.5	19.4	164.7
Rice-Rice+Fish+75%N	12.4	6.8	7.5	0	0	26.7
Rice-Rice+Azolla+Fish+75%N	16.5	8.0	8.3	104.7	0	137.5
Rice-Rice+Fish+GLM+75%N	13.4	8.2	11.1	0	19.4	52.1
Rice-Rice+ Azolla + Fish + GLM+75%N	17.0	9.1	11.2	106.3	19.4	163.0

(Balusamy, 1996)

uplands are Sorghum - cotton - maize, Ragi - cotton - maize and Ragi - cotton - sorghum. The various components integrated were dairy, biogas, spawn and mushroom production, apiary, homestead garden, fish rearing and silviculture.

Integrated farming system involving crop, dairy, biogas and silviculture was taken from 1987-92. The normal cropping pattern followed in Coimbatore is ragi-cotton-sorghum. In the integrated approach, the same cropping pattern was slightly modified by inclusion of intercropping such as greengram in cotton, cowpea in sorghum and sunflower as border crop in ragi (0.75 ha). The perennial grass fodder (0.15 ha) and legume fodder lucerne (0.05 ha) were also raised in the holding.

Three jersey cross bred milch cows with 2 calves were included under dairy component. For effective recycling of farm and animal waste, a bio-gas unit of 2 m^3 capacity was installed for the production of fuel, light and enriched manure (Plate 6). Sixty kg of cowdung expected out of 5 animals is sufficient enough to produce 2 m^3 of gas everyday which is equivalent to 1.5 litres kerosene. Two hundred numbers of subabul tress were planted all along the boundary of the field for fodder and timber production. The results of the study revealed that the entire system produced a net income of Rs.20702 ha^{-1} $year^{-1}$ (Models 9 and 10). This system also facilitated effective recycling of farm and animal waste, improved farm employment opportunities and continuous flow of income to farm throughout the year.

By this recycling, some of the weed seeds present in the raw cow dung also get killed during digestion process, thus improving the quality of the slurry over its raw material used *viz.*, cow dung. The quality improvement by way of recycling the cow dung through biogas chamber was studied by analyzing the NPK content of slurry and FYM

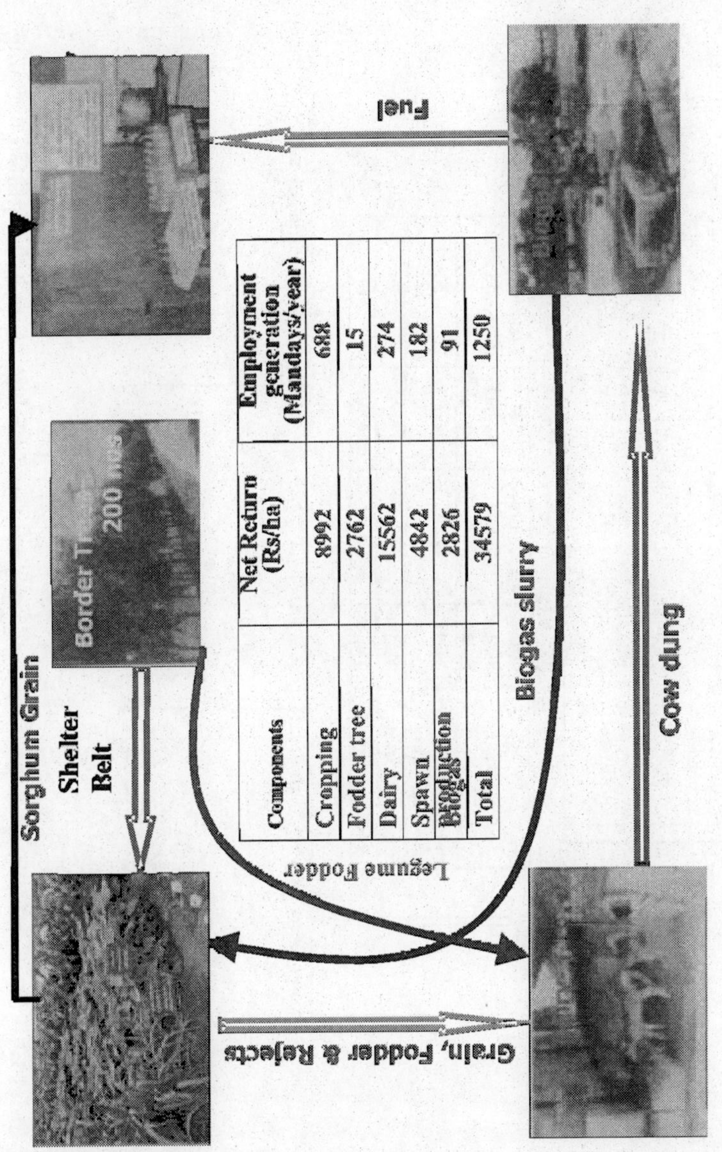

Components	Net Return (Rs/ha)	Employment generation (Mandays/year)
Cropping	8992	688
Fodder tree	2762	15
Dairy	15562	274
Spawn production	4842	182
Biogas	2826	91
Total	34579	1250

Model 9. Integrated Farming System for Irrigated Upland Ecosystem (1 ha)

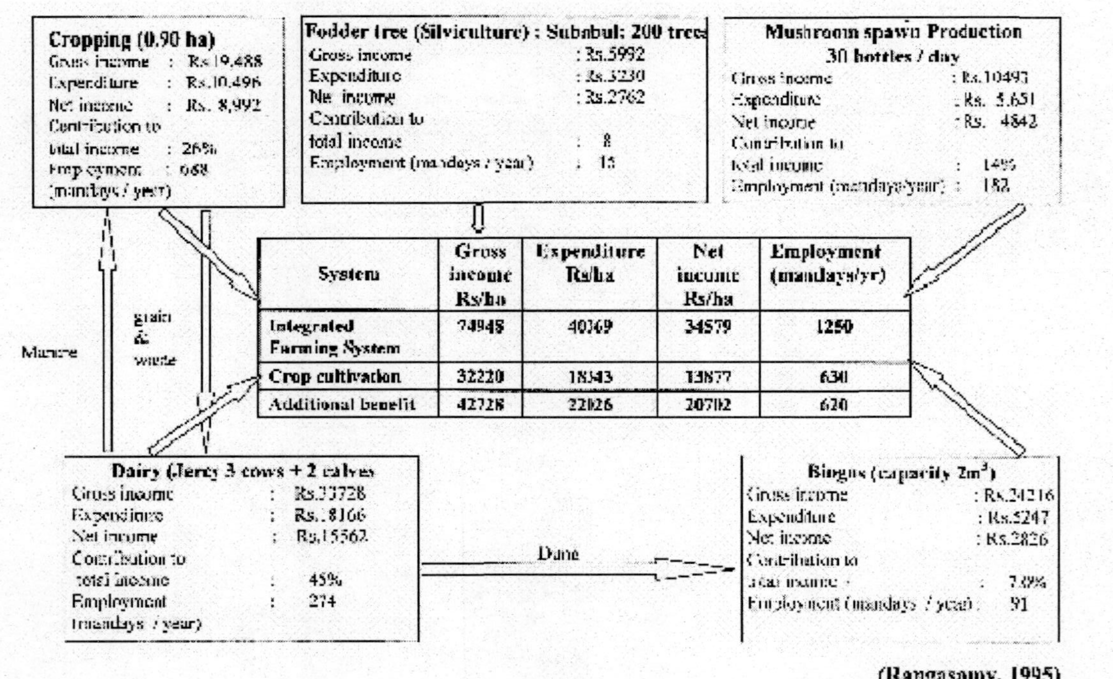

Model 10. Resource Flow in Crop + Dairy + Biogas + Spawn + Silviculture in Integrated Farming

prepared utilizing the cow dung from the system. The total quantity secured from the unit over the period of 365 days has been taken on equal weight basis and the analyzed data is furnished in the Table 13 and Table 14.

Recycling of cow dung lead to the production of 730 m^3 of biogas with the possibility of enhancing the nutrient value of NPK to the tune of 44.5 kg, 65.9 kg and 28 kg respectively, in a year. Trace elements like Fe, Mn, Zn and Cu are also present in an enhanced level over FYM (Rangasamy, 2000).

Table 13. Nutrient enhancement through biogas slurry and FYM

Particulars	Major element (%)			Trace element (ppm)			
	N	P	K	Fe	Mn	Zn	Cu
Biogas slurry	1.43	1.21	1.01	4200	550	150	52
Farm yard manure	0.94	0.56	0.72	4000	490	100	45

Table 14. Nutrient gain by recycling (kg)

Particulars	N	P	K	Fe	Mn	Zn	Cu
Nutrient in biogas slurry (11.0 t on dry weight basis)	157.3	133.1	144.4	46.2	6.05	1.65	0.57
Nutrients in FYM (11.6 t) in an year	112.8	67.2	86.4	44.0	5.39	1.10	0.49
Nutrients gained by way of recycling	44.5	65.9	28.0	2.2	0.66	0.55	0.08

(Rangasamy, 2000)

Crop + Dairy + Biogas + Mushroom + Fish

In this system components of integrated farming system involved were crop + dairy, crop + dairy + biogas + fish, crop + dairy + biogas + mushroom + fish and crop + fish (Plate 7).

Experimental results on farming systems for irrigated uplands revealed that integration of cropping with components like dairy, biogas and mushroom or fish resulted in higher productivity than cropping alone. During 1999-2000, integration of 0.32 ha each of sorghum + red gram - sunflower + coriander, maize + fodder cowpea - cotton + coriander and perennial fodder CO 3 grass + legume fodder (Lucerne) with dairy (6 cows + 4 calves), biogas (2 m^3 capacity), mushroom (2 kg day^{-1}) + spawn production (10 bottles day^{-1}) and fish (0.008 ha pond with 80 fingerlings fed by cattle shed washing) yielded higher sorghum productivity than cropping alone with sorghum - cotton (0.50 ha) and maize - cotton (0.50 ha) cropping systems. During 2000-2001, integration of 0.32 ha each of sorghum + red gram - sunflower + coriander, maize + fodder cowpea - cotton + coriander and perennial fodder CO3 grass + legume fodder (lucerne) with dairy (6 cows + 4 calves), biogas (2 m^3 capacity) and mushroom (2 kg day^{-1}) + spawn production (10 bottles day^{-1}) recorded higher productivity than the cropping alone with sorghum - cotton (0.50 ha) and maize - cotton (0.50 ha) cropping systems (Table 15).

Cropping + dairy + biogas + mushroom + fish (cattle shed washing) integration earned the highest gross and net returns with better per day returns. These farming systems registered the benefit - cost ratio of 2.22 during 1999-2000. Cropping + dairy + biogas + fish + mushroom integration recorded the highest gross, net and per day returns. It also registered the highest benefit cost ratio of 2.41 during 2000-2001 (Table 16).

Integration of cropping with dairy + biogas + mushroom generated the highest employment of 875 man days. The allied enterprises generated additional employment of 469 man days providing opportunity for 1.28 family members to be employed round the year (Table 17) (Sivamurugan, 2001).

Table 15. Productivity (kg) of components of integrated farming systems

Farming systems	1999-2000						2000 - 2001					
	Crop	Dairy	Biogas	Fish	Mushroom	System productivity (kg ha^{-1})	Crop	Dairy	Biogas	Fish	Mushroom	System productivity (kg ha^{-1})
FS_1 Cropping alone	10335	-	-	-	-	10335	11091	-	-	-	-	11091
FS_2 Crop (FYM) + Dairy	12160	20598	-	-	-	32778	12622	21102	-	-	-	33924
FS_3 Crop (BGS) + Dairy + Biogas + Fish (BGS)	11945	20598	1168	227	-	33938	11980	21102	1168	245	-	34495
FS_4 Crop (MSS) + Dairy+ Biogas + Mushroom + Fish (CSW)	11643	20598	1168	245	11196	44545	12227	21102	1168	-	11483	45980
FS_5 Crop + Fish (AF)	11443	-	-	236	-	11679	11813	-	-	255	-	12068

(Sivamurgan, 2001)

* Sorghum grain equivalent yield

FS_1 - Sorghum - cotton - 0.50 ha

Maize - cotton - 0.50 ha

FS_2 to FS_5 - Sorghum + red gram - sunflower + coriander - 0.32 ha

Maize+ fodder cowpea cotton + coriander - 0.32 ha

Perennial fodder CO.3 grass + legume fodder (Lucerne) - 0.32 ha.

Fish (CSW) - Fish fed with cattle shed washing

FYM - Farmyard manure

BGS - Biogas slurry

MSS-Composted mushroom spent substrate

AF - Fish pond silt from the pond fed with artificial feeding

Fish (BGS)-Fish fed with biogas slurry

Fish (AF) - Fish fed with artificial feed

Table 16. Economic analysis (Rs.) of individual components in integrated farming systems

Components	1999-2000			2000-2001		
	Cost of production	Gross return	Net return	Cost of production	Gross return	Net return
Cropping alone - 1.00 ha	31525	62010	30485	31525	66546	35021
Crop with FYM - 0.96 ha	30339	70722	40383	30339	74448	44109
Crop with BGS - 0.96 ha	29174	69996	40822	29174	73356	44182
Crop with MSS - 0.96 ha	28854	68712	39858	28854	71574	42720
Crop with AF - 0.96 ha	28774	66444	37670	28774	68592	39818
Dairy (6 cows + 4 calves)	62382	126590	64208	28774	70200	41426
Biogas (2m^3 capacity)	6764	15753	8989	28774	69516	40742
Mushroom (2 kg day^{-1}) + spawn production (10 bottles day^{-1})	27891	67872	39981	57252	130993	73741

Contd....

Table 16. Contd....

Fish (0.008 ha pond)						
Artificial feeding	790	1575	785	783	1687	904
Biogas slurry	446	1521	1075	446	1630	1184
Cattle shed washing	461	1619	1158	441	1916	1475

(Sivamurugan, 2001)

Table 17. Employment generation of integrated farming systems

Farming systems	Employment generation (man days)					
	Crop	Dairy	Biogas	Fish	Mushroom	Total man days
FS_1 Cropping alone	394	-	-	-	-	394
FS_2 Crop (FYM) + Dairy	406	286	-	-	-	692
FS_3 Crop (BGS) + Dairy +Biogas	406	286	12	-	-	704
FS_4 Crop (MSS) + Dairy +Biogas +Mushroom	406	286	12	-	171	875
FS_5 Crop (AF) + Fish (AF)	406	-	-	8	-	414
FS_6 Crop (BGS) + Fish (BGS) +Dairy + Biogas	406	286	12	1	-	705
FS_7 Crop (CSW) + Fish (CSW) +Dairy	406	286	-	3	-	695

(Sivamurugan, 2001)

4.1.3 Rainfed land

The meteorological data of different agroclimatic regions of Tamil Nadu have clearly indicated that the seasonal rainfall in rainfed areas is very low and the distribution is also highly erratic. It is evident that if the required moisture and nutrient at the critical growth phase of the short duration field crops are not satisfied, the yield of the crop will be affected drastically. Most of the years, the farmer experiences complete failure of the crop due to non-availability of moisture at the critical stage. This is the reason why conventional rainfed agriculture is said to be out and out a gamble.

To overcome complete failure in the rainfed areas through traditional crop activity being practiced, integrating different enterprises and utilising the biomass built up have been identified as a successful venture to give regular income to the rainfed farmers. There is a good scope for getting required biomass even with the available erratic seasonal rainfall, to the allied enterprises likely to be integrated *viz.*, goat, buffalo, pigeon, rabbit, etc. The outcome of these enterprises will be an alternate source for protein, CHO-Carbohydrate, fat, minerals, vitamins and energy. Drought tolerant perennial forest wood /timber value trees can also be raised utilising the rainfall received round the year and can be a good source for valuable fuel wood or timber after some years (Plate 8). Similarly, drought tolerant perennial horticultural fruit crops can also be raised utilising the rainfall received round the year and thus could help in improving the income of the resource poor rainfed farmers. The integrated farming system when implemented properly in the rainfed areas throughout the country, the present contribution from 2/3 net cultivated rainfed land *viz.*, 42 per cent can be doubled without any difficulty.

The linkage of other enterprises *viz.*, goat, buffalo, pigeon. rabbit etc will also provide good amount of organic source of nutrients to the soil. As we well aware, rainfed soils are not only thirsty but also hungry. If we could improve the nutrient status of the existing rainfed soils through said linkages; it could enhance the yield substantially, with the prevailing conditions of the rainfed tract. Considering all these factors, Integrated farming system models consisting of varied enterprises have been developed at Tamil Nadu Agricultural University, Coimbatore.

Crop + goat

An integrated farming system study involving grain crop, fodder crop, fodder trees, perennial grasses and goat rearing in an area of one hectare of rainfed land was carried out at Coimbatore from 1987-1992 (Plate 9). Generally, in dryland farmers raise only fodder sorghum yielding 3 to 10 tonnes of fodder/ha depending upon the rainfall. In the integrated approach, the cropping pattern was modified by including both sorghum grain crop (0.2 ha) and sorghum fodder crop (0.2 ha). To meet the fodder requirement of goats, 0.2 ha of land was raised with subabul and *Cenchrus ciliaris*, a pasture grass as an intercrop. Trees like *Acacia senegal* and *Prosopis cineraria* were raised in 0.20 ha. The goat unit comprised of Tellichery goats 20 eves and 1 buck.

The results revealed that through short duration field crops and perennial crops the feed requirement for one productive unit consisting of 20 eves and one buck for all the 365 days can be met. After 5 years, the perennial fodder trees could bear the shock and would yield sufficient quantum of loppings to supplement the feed requirement of 21 adults along with millets, legumes and perennial grass linked. The Tellicherry goat will be a good breed to build up body weight for every unit of feed secured through different sources under rainfed condition. It is a dual purpose animal, where it gives 80 to 100 ml of milk after satisfying the full requirement of dependent kids on it every day. Twenty productive females could give 45 kids per annum. Each kid at the time of weaning will weigh around

12 kg. Moreover the unit of 21 animals with different stages of kids under deep litter system would give 11.2 t of valuable manure. This, when applied to the soil, will not only an excellent source of primary, secondary and micro nutrients for the crops but also absorb more moisture, retain in the soil and releases to the crop appropriately for better yield.

The net income from the farming system was Rs.5671 ha^{-1} $year^{-1}$ and that of control Rs.1919 ha^{-1} $year^{-1}$. Out of the total income from the integrated farming systems, 59 per cent was from goat rearing. The additional net income realized from integrated farming system was Rs.3752 ha^{-1} $year^{-1}$ as compared to cropping alone. The additional employment gained through integrated farming system over cropping was 314 man days ha^{-1} $year^{-1}$ (Models 11 and 12).

The organic manures like, litter from the goat unit can readily be used for soil application, and thus will help in enriching the soil. Goat droppings are found to be a good energy source, which can also be linked with biogas unit before it is utilised as manure. This will generate good volume of gas (22 kg of goat dropping will generate one cubic meter of gas as against 30 kg of cattle dung) as well as enhance nutrient availability. Thus, through recycling of organic in the Farming Systems approach, the potential of each produce can be exploited to a greater extent. The data collected on the availability of organic source of nutrient through 20 + 1 goat reared under deep litter system are presented below (Rangasamy, 1995).

Table 18. Nutrient gain from 20+1 productive Tellichery goats unit under deep litter system

No. of Kidding $year^{-1}$	: 1.5
No. of Kids $Kidding^{-1}$	: 1.5
No. of Kids $year^{-1}$ $adult^{-1}$	: 2.25

Contd...

Table 18. Contd...

No. of Kids from 20 adults year^{-1}	: 45
Mean weight of dropping day^{-1}	: Adult : 900g Kid : 110g
Mean weight of dropping year^{-1}	: 8600 kg
Coir waste used for the stall year^{-1}	: 2600 kg
Deep litter waste obtained year^{-1}	: 11200 kg (8600 + 2600)
Nutrient content of deep litter waste (%)	: 1.75 N : 0.95 P: 0.82 K
Total nutrient (kg) available from the goat unit year^{-1} (kg)	: 200 N : 106 P : 91K

Crop + Tree + Goat

Another integrated farming system model involving crop + tree + goat was taken from March 1999 to February 2001. Conventional cropping system with sole sorghum was compared with *Ailanthus excelsa* + crop + goat, *Ceiba pentandra* + crop + goat and *Emblica officinalis* + crop + goat to identify the suitable component linkage. The cropping systems included were grain sorghum + cowpea, fodder sorghum + cowpea and *Cenchrus glaucus* each in 0.33 ha in integrated farming systems and the remaining 0.01 ha was allotted to the goat component. One unit of Tellichery goats, consisting five females and one male was included in the system. Three tree species *viz., A. excelsa, C. pentandra* and *E. officinalis* were evaluated for their performance under dryland situation to evaluate the suitable *in situ* moisture conservation practices and nitrogen management on yield of crops in the integrated farming system.

Experimental results on integrated farming system revealed that (i) integration of sorghum + cowpea (grain), sorghum + cowpea (fodder) and *C. glaucus* each in 0.33 ha intercropped in *E. officinalis* with Tellichery goat component

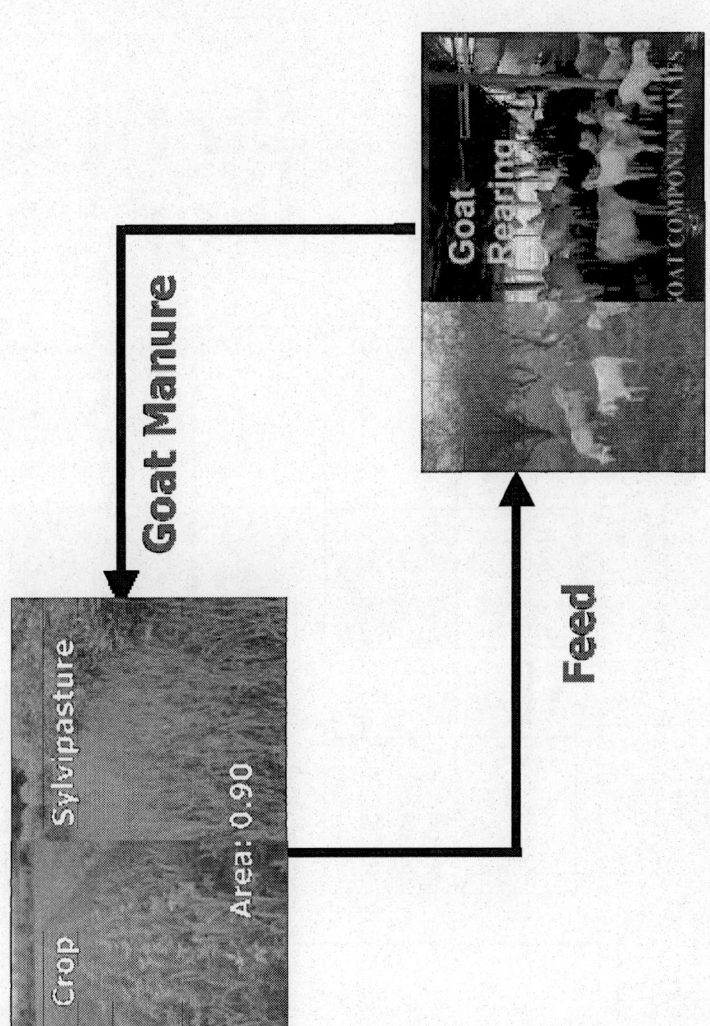

Model 11. Component Linkage in Integrated Farming System for Rainfed Land

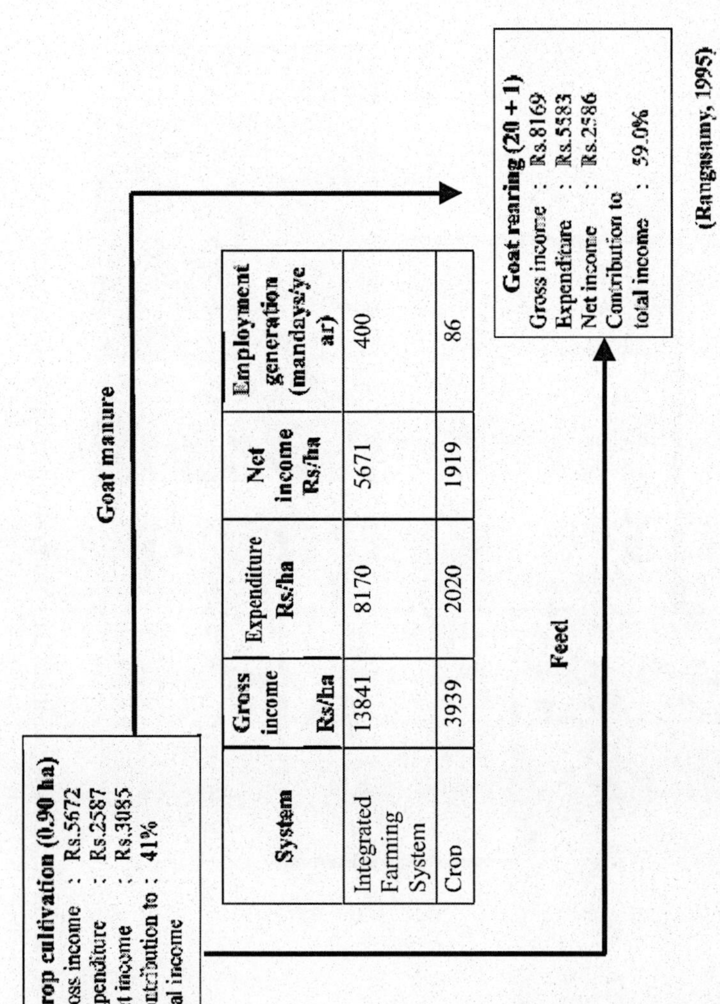

System	Gross income Rs/ha	Expenditure Rs/ha	Net income Rs/ha	Employment generation (mandays/year)
Integrated Farming System	13841	8170	5671	400
Crop	3939	2020	1919	86

(Rangasamy, 1995)

Model 12. Resource Recycling in Crop + Goat Integrated F arming System

(5+1) in 0.01 ha resulted in higher productivity, economic returns and provided better employment opportunity and improved soil fertility than raising sole sorghum alone (Table 19, 20 & 21) (ii) growth of *E. officinalis* was better as compared to other trees under *vertisol* dryland situation, (iii) coir pith mulching and pitcher irrigation increased the tree seedling growth than the control, (iv) tied ridges conserved more moisture and improved the productivity of the crops, (v) application of 50 per cent N through fertilizer and 50 per cent N through goat manure increased the productivity, enhanced the soil fertility and provided better opportunity for recycling of manure to the crops (Radhamani, 2001).

Integration of various agricultural enterprises *viz.*, cropping, forestry, animal husbandry, fishery, etc., has great potentialities in the agricultural economy. These enterprises not only supplement the income of the farmer but also help in increasing the family labour employment. Inclusion of livestock enterprise in farming system allows for more effective use of natural resources like climate, land, soil and vegetation. Livestock converts weeds, crop residues and by-products into food stuffs for human consumption, produces milk and meat and provides power in the farm besides its manure which improves the soil productivity. Multipurpose forest trees and forage crops grown in rotation and with agricultural crops provide feed to animals besides controlling erosion and conserving soil productivity. These enterprises stabilise seasonal and annual food production, labour and power requirements and increase net farm income.

The future of Indian agriculture depends heavily on the development of appropriate farming system as it fits to resource poor farm families of different agro ecological zones. The endowment of abundant sunshine, long growing season, responsive soil type and combination of surface water, ground water and seasonal rains and above

Table 19. Productivity contribution of components of integrated farming systems

Farming systems (1.00 ha)		*Component productivity (tonnes)								
		1999			2000			Mean		
		Crop	Goat	Total	Crop	Goat	Total	Crop	Goat	Total
Conventional cropping		9.09	-	9.09	3.59	-	3.59	6.34	-	6.34
Ailanthus excelsa + crop + goat	CS_1	9.79			3.31			6.55		
	CS_2	3.93			2.54			3.24		
	CS_3	3.76			3.14			3.45		
	Goat	-	14.1		-	22.6		-	18.35	
	Total	**17.48**	**14.1**	**31.58**	**8.99**	**22.6**	**31.59**	**13.24**	**18.35**	**31.59**
Ceiba pentandra + crop + goat	CS_1	8.62			3.41			6.02		
	CS_2	3.80			2.50			3.15		
	CS_3	3.74			3.12			3.43		
	Goat	-	14.1		-	22.6		-	18.35	
	Total	**16.16**	**14.1**	**30.26**	**9.03**	**22.6**	**31.63**	**12.60**	**18.35**	**30.95**

Contd...

Table 19. Contd...

	CS_1	9.78			3.83			6.81		
	CS_2	4.01			2.97			3.49		
Emblica officinalis + crop + goat	CS_3	3.74			3.15			3.45		
	Goat		14.1		-	22.6		-	18.35	
	Total	**17.53**	**14.1**	**31.63**	**9.95**	**22.6**	**32.55**	**13.75**	**18.35**	**32.10**

* Sorghum fodder equivalent

CS_1 - Grain sorghum + cowpea - 0.33 ha
CS_2 - Fodder sorghum + cowpea - 0.33 ha
CS_3 - *Cenchrus glaucus* - 0.33 ha
Goat - 5 + 1 - 0.01 ha

Table 20. Economic analysis of integrated farming system components

Farming systems	1999				2000			
	Cost of production (Rs. ha^{-1})	Gross return (Rs. ha^{-1})	Net return (Rs. ha^{-1})	B : C ratio	Cost of production (Rs. ha^{-1})	Gross return (Rs. ha^{-1})	Net return (Rs. ha^{-1})	B : C ratio
Conventional cropping	3618	3816	198	1.05	3618	1509	-2109	0.42
A. excelsa + crop + goat	17379	20450	3071	1.18	15393	16306	913	1.06
C. pentandra + crop + goat	17379	19898	2519	1.14	15393	16315	922	1.06
E. officinalis + crop + goat	17379	20470	3091	1.18	15393	16536	1143	1.07

Table 21. Employment generation (man days) of integrated farming system

Farming systems	1999				2000			
	Crop	Tree	Goat	Total	Crop	Tree	Goat	Total
Conventional cropping	37	-	-	37	37	-	-	37
A. excelsa + crop + goat	44	17	23	84	40	17	35	92
C. pentandra + crop + goat	44	17	23	84	40	17	35	92
E. officinalis + crop + goat	44	17	23	84	40	17	35	92

(Radhamani, 2001)

all a progressive peasantry offer vast scope for an intensive farming system through multiple cropping and diversified farming including animal husbandry, forestry, sericulture, fisheries and the like. The challenge is to upgrade technological and social disciplines on a continuous basis and integrate these disciplines to suit the region and the farm families in a manner that may ensure increased production with stability, ecological sustainability and equitability.

Thus the integrated farming system approach introduces a change in the farming techniques for maximum production in the cropping pattern and takes care of optimal utilization of resources. The farm wastes are better recycled for productive purpose in the integrated system. Farm as a unit is to be considered and planned for effective integration of all enterprises.

Research studies carried out in lowlands, uplands with supplemental irrigation and rainfed lands have demonstrated the technical feasibility and economic viability of the integrated farming systems attempted. Besides facilitating cash income, these models generate additional employment for family labour and minimize the risks associated with mere crop production.

The results also clearly show that the recycled organic residues / wastes of crops and allied activities could supplement the chemical fertilizers. Moreover, this will also provide opportunity to reduce the environmental pollution by way of burning the fossil fuel in the process of manufacturing the chemicals. The soil health problem by way of dumping more and more quantities of inorganic can also be solved to greater extent by proper utilization of organics under different field situations.

Questions

1. Define farming system scope, need and principle
2. Give the possible components and possible integration in integrated farming system for different ecosystem
3. Explain fish rearing and mushroom cultivation in integrated farming system
4. Give in detail the advantages and constraints of dairy farming and goat rearing in integrated farming system
5. Interaction between rice cum fish culture, crop + livestock and crop + poultry in integrated farming system
6. Define agroforestry and their role in integrated farming system
7. Factors influencing choice and size of enterprises in integrated farming system
8. Enlist advantages of integrated farming system
9. Give an integrated farming system model for wetland or irrigated upland ecosystem and show the resource recycling potential

10. Integrated farming system -A way to organic agriculture - Substantiate
11. Integrated farming system - A way to sustainable agriculture - Substantiate
12. Write about farming system research and evaluation methods

—x—

Plates

Plate 1. Pigeon Component

Poultry + Fish Integration

Plate 2. Pigeon + Fish Integration

Plate 3. Goat + Fish

Plate 4. Fish Component

Plate 5. Rice + Fish + Azolla System

Plate 6. Diary + Biogas Integration

Plate 7. Spawn + Mushroom Integration

Plate 8. Border Trees + Farm Pond

Plate 9. Agroforestry

ABOUT THE AUTHORS

Dr. C. Jayanthi, Ph. D., Professor, Department of Agronomy, Tamil Nadu Agricultural University, has 21 years of teaching, research and extension experience in Farming system Research and Forage crops. She completed her B. Sc. (Ag.) during 1983 and M. Sc. (Ag.) during 1985. She obtained Ph. D during 1995 from Tamil Nadu Agricultural University, Coimbatore-3. She is a recipient of FAI Silver Jubilee award and Indian Society of Agronomy award for best research on Integrated farming system, Best paper presentation prize from Indian Society of Agronomy, Best Researcher Award of Tamil Nadu Agricultural University, Best Research Paper Award from Madras Agriculture Students Union, and Best Research Paper Award for research article on farming system from Indian Society of Agronomy. She has 5 books, 9 booklets and technical manuals, 62 research articles and 50 popular articles to her credit. Participated in three months training on nutrient management in rice based cropping system at IRRI, Manila, Philippines. She has attended International symposium on Farming system research organized by Asian Farming Systems Association and FAO during 1994, at Manila, Philippines and presented a research paper on Farming system. She has attended several trainings, symposium, seminars and conferences at national and international level. She has served as a resource person for many training and conference on forage crops and farming system at state and national level. She has a rich research experience on farming system by heading two research projects as Principal investigator. She has guided three post graduate students and one Ph. D. student. She is an active member in many professional bodies like Indian Society of Agronomy, Asian Farming Systems Association, Madras Agricultural Journal, Journal of Farming Systems Research and Development and Kisan World. Presently she is an Editorial board member in Indian Society of Agronomy. She is one of the projects screening committee member nominated by Indian Council of Agricultural Research for farming system research projects

Dr. P. Devasenapathy, Ph. D., Professor, Department of Agronomy Tamil Nadu Agricultural University, has 25 years of teaching, research and extension experience. He completed his B. Sc. (Ag.) during 1979 and M. Sc. (Ag.) during 1981. He obtained Ph.D during 1994 from Tamil Nadu Agricultural University,

Coimbatore-3 with specialization in cropping system research and soil fertility. He worked in various disciplines like Water Technology Centre, Training Division and Energy Requirement Scheme. Presently working as Chief Agronomist of the AICRP on cropping systems and Principle Investigator for Organic Farming Scheme at Tamil Nadu Agricultural University. Visited IRRI, Philippines under AHRD-FAO fellowship for training on nutrient management in rice based cropping system and Srilanka for International conference. He has 60 research papers in both national and international journals to his credit and guiding Post Graduate and Ph. D. students of Agronomy. He has participated in several trainings, symposium, seminars and conferences at national and international level. He has handled World Bank projects and coordinated for the development of policy documents to Government of Tamil Nadu.

Dr. C. Vennila, working as Research fellow in the Scheme "Farming System Research". She completed her B. Sc. (Ag.) during 2001 and M. Sc. (Ag.) during 2003. She obtained Ph. D during 2006 from Tamil Nadu Agricultural University, Coimbatore-3 with specialization in organic farming. She is a recipient of Jalgoan Jain Irrigations Award, Ramasamy Award, Daniel sundarraj Medal, Sathiyavan Gopala Pillai Award and Dr. P.R. Mehta Award for undergraduate programme and Best Student award for her M.Sc. (Ag.,) programme. She has published two research articles and three popular articles in National and International journals. She has attended 4 National and 2 International Conferences and Seminars. Associated the courses on Principles of Crop Production, Integrated Farming System, Principles of weed science, irrigation and dry farming management for under graduate degree programme.